Electrical Repairs

BY WILLIAM BERNARD

GROSSET
GOOD LIFE
BOOKS

PUBLISHERS • GROSSET & DUNLAP • NEW YORK

Acknowledgments

Cover photograph by Mort Engel

The author wishes to express his appreciation to the following organizations for permission to use their illustrations in this book: Cable Electric Products, Inc.; Challenger Safe Home Alarm System; Departments of the Army and Air Force; Eagle Electric Mfg. Co., Inc.; General Electric Company; Hartman Electric Co., Inc.; Knight Watch Systems, Inc.; Leviton Mfg. Co., Inc.; Montgomery Ward & Co.; Scoville (Nutone Division); Pass & Seymour, Inc.; Sears, Roebuck & Co.; Sylvania; Swingline, Inc.; 3M Company. Do-it-yourself photographs and those showing fuse and wiring details are by Robert H. Feuchter.

Illustrations on pages 14, 15, 35, 54 reprinted by permission of Home Pro Electrical Installation and Repair Guide. Copyright © 1975 by Minnesota Mining and Manufacturing Company. All rights reserved.

Contents

1
How the System Works

As far as the do-it-yourselfer is concerned, the electric current from the utility company enters your home through a metal box. This box may be located in your basement, in your kitchen, or even in a closet or some other hidden place — but there is always a box. It may be large or small. It is variously called a *service entrance*, a *service panel*, a *fuse panel*, or simply a *fuse box*. Usually the entrance contains fuses — hence the term fuse box. Sometimes the entrance contains circuit breakers — automatic switches that do the work of fuses. Let's first consider fuse boxes.

The fuse box keeps your home from catching fire when an overload or a short circuit causes electrical wires to get too hot. To repair any fault in wiring, outlets, switches, or light fixtures, you must go first to the fuse box.

Fuse boxes come in many shapes and varieties. The most common are fused entrances that usually receive two lines that enter your house from underground or from utility poles on the street. One line is *hot* and delivers from 110 to 120 volts — usually about 115 — of electricity. The other is the so-called *neutral* or *ground* line. Within the box, the current is fed to branch lines or circuits serving the various lights and appliance outlets. Each branch circuit is protected by a small round fuse.

The box also contains (or is connected to) a main switch, with which you can cut off all electricity. Often the switch is activated by a hefty handle on the outside of the box. Sometimes the handle lies inside. There are also boxes with doors that mechanically cut off all the current when they are opened.

Modern Fuse Boxes

In addition to the familiar round fuses, more modern fuse boxes contain one or more blocks or "drawers," each with recessed handles. Instead of using a switch to cut off the inflow of all current, you do so by pulling a block out of the box. On the back of the block (or within it) is a different kind of fuse. It has the shape of a tube or a cartridge.

If you live in an apartment, it's likely that the main switch or block is located in the basement of your building. You cannot shut off the current from your apartment. However, your apartment should have a box containing two or more round fuses.

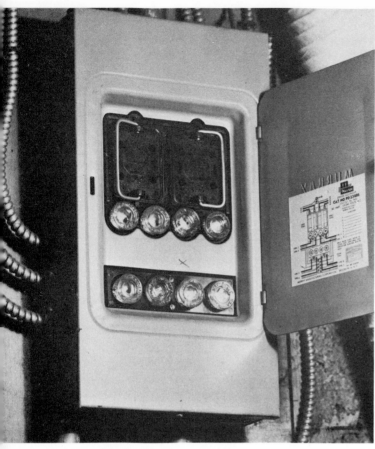

Standard fuse box with screw-in fuses. Each fuse is a safety device protecting an individual circuit. Block at upper left is main switch. Block at upper right is range switch.

Some fuse boxes have takeoff circuits for special purposes. Photo shows takeoff (for dryer) guarded by its own large-amperage fuse. At lower left is a utility outlet on another takeoff.

Because of the growing demand for electricity, another kind of fuse box is widely seen in homes today. This type receives three lines — two hot and one neutral. Since each hot line has a potential of some 115 volts, and since both lines can operate via the single neutral line, a greater number of 115-volt branch lines can be used. Or, the two lines can work together to deliver 230 volts — as required for an appliance such as an electric range.

Fuse boxes serving these three-line systems nearly always contain at least two blocks. The main block sits astride all the 115-volt branch lines; when the block is pulled out, none of the branch circuits will operate. The second, or "range," block guards the 230-volt circuit; when this block is pulled out, the circuit is dead.

If a house has no 230-volt circuit but its fuse box contains two or more blocks, each protects a separate system of 115-volt branch lines.

Occasionally a takeoff from the fuse box leads to a separate smaller box that also contains fuses. The fuses guard one or more branch circuits that were probably installed at a later date than the larger original fuse box.

Utility lines go first to a meter before entering the fuse box. The meter tells how many kilowatt-hours your home used during a month or other billing period, and the utility company bills you accordingly. In two-line systems, the meter is usually inside the house. Meters for three-line systems generally are located outside the house.

Fuses for Failsafe

Fuses must be used in residential and commercial electrical circuits to safeguard them from overloads and shorts.

The electrical capacity of a circuit depends on the size and length of the wire, the insulation, and other factors. When a circuit's capacity is exceeded, the wire heats up unduly. The extra heat travels to a fuse, melting the special metal within it. This creates a gap in the circuit, automatically shutting it off.

If electrical circuits were not protected by fuses, an overloaded or shorted circuit would grow hotter and hotter — ultimately burning up the wiring and maybe setting your house afire.

The common round fuses you see in the box come marked with their amperage ratings. Most household circuits take a 15-amp fuse. Circuits designed for somewhat greater capacity — for instance, to serve the plug-in appliances in a kitchen — may take a 20-amp fuse. Circuits that carry 230 volts (such as for an electric range) or serve other heavy-duty equipment (such as a dryer or power tools) require a 30-amp fuse. As for a 30-amp fuse, it should never be allowed to get into your fuse box at all — unless your home has a special heavy-duty circuit for electrical tools, motors, air conditioners, or the like.

Never substitute a higher-rated fuse for a lower-rated fuse. For example, you can make a circuit work by screwing a 20-amp fuse into a socket for a 15-amp fuse, but you are defeating the purpose of the fuse. A fuse rated too high for the capacity of the circuit will not shut off a shorted or overloaded circuit soon enough. It will permit heat to build up beyond the danger point — with possibly disastrous results.

Fuses often come packaged. These are the tamperproof type with special bases.

Special Fuses

Type-S fuses have been developed to make sure that a blown fuse is not replaced by a fuse of the wrong rating. Sold commercially under names like Non-Tamp and Fustat, type-S fuses look and act like any ordinary fuse. However, the bases of these fuses come in different sizes — one size is for 15-amp fuses, another size is for 20-amp fuses, and still another size is for 30-amp fuses. Each base has its own adapter ring.

Before you can use type-S fuses in your home, you must insert the appropriate adapter rings into the fuse sockets. Begin by screwing the right-sized fuse into the adapter ring. Then screw the fuse and the adapter together into the fuse socket. The adapter will lock permanently into the socket. You can still unscrew a blown fuse, but you can replace it only with one having the same-size base — and therefore the correct rating.

Another special type of fuse is the *time-delay* or *slow-blow fuse*. This type of fuse responds

Time-delay fuse does not blow under temporary overload.

Cartridge fuse. These are for heavy-duty, as in main switches. If they blow, the "hot" side retains the temperature for some time.

Mounted there is at least one *cartridge fuse,* so called because of its shape. Such fuses are customarily used as main fuses in homes and apartment buildings — protecting the main circuit just as the smaller screw-in fuses protect the branch circuits.

Main fuses are rated from 30 amperes on up to 100 amperes or more. Their job, primarily, is to protect homes from catastrophic power surges in the utility company's power lines. For example, a lightning bolt striking a transmission cable could blow your main fuse.

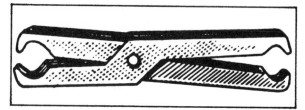

Fuse-pulling tool to handle cartridge-type fuses.

When a main fuse does blow, not a light or outlet in your house will work. Replace the fuse with another one of exactly the same size, type, and amperage rating. If your house remains dark, the trouble is in the utility company's lines. Check neighboring houses. If they are dark too, the trouble extends over your area. If they have electricity, the trouble is in the lines from the utility poles to your house. Telephone the utility company and report your problem.

Fuse boxes without blocks also depend on a main fuse to protect the house wiring from outside power surges. In such boxes the cartridge-type main fuse is hooked up to the main switch.

Fortunately, there is almost no chance that the main fuse will need to be changed. Should that contingency arise — as a result, say, of an electrical storm — most likely the utility company will send experienced repairmen into your neighborhood. They will change the fuse for you.

Although they are supposed to be covered, the connections may be bare — so be careful not to touch any exposed wire or terminals.

If, for some reason, you must replace a main fuse yourself, make certain that the main switch is in the *off* position before you begin and that you are not standing on a wet spot.

immediately to a short circuit, but if an overload occurs, two or three seconds elapse before the fuse melts.

These slow-blow fuses were developed to handle demands put on circuits by electric motors. When an electric motor is switched on, it may require as much as five times more electricity to get itself into motion than it does while actually running. The surge of extra power may be enough to blow an ordinary fuse, but the slow-blow fuse holds out long enough for the surge to pass. If your air conditioner or some other motored appliance is in the habit of blowing fuses, you might be able to cure the condition by installing a slow-blow fuse.

A third type of special fuse is one you will see if the main switch of your fuse box is of the block-type. Grasp the handle, pull out the block, and then look at its rear or interior.

Wear thick rubber gloves. If possible, keep one hand in your pocket, so that it will not accidentally contact bare metal somewhere.

Circuit Breakers

Although most service entrances do indeed contain fuses, some contain instead two to a dozen or more automatic switching devices. Automatic, that is, in switching off current when a circuit gets too hot. They do not automatically turn the current back on. After a circuit has been turned off, the device must be reset before the current can flow again — however, resetting takes only a push of your finger. These handy devices are called *circuit breakers*.

Circuit breakers eliminate the need to replace fuses. They are neat and reliable. They are safe to handle because, once installed, no

To kill the current in a branch circuit, flip the breaker to the off *position. To restart the current, flip the breaker to* on.

metal parts are exposed. They operate on a slow-blow principle, accurately compensating for current surges.

As a result, the trend in modern homes is toward service entrances that contain circuit breakers rather than fuses. In addition, many older residences have been fitted with circuit-breaker boxes replacing fuse-bearing ones. Circuit breakers are manufactured in the same ampere ratings as cartridge fuses and the lighter-duty round fuses. Thus, breakers may replace the main switch and main fuse, the range switch and range fuse, and every other switch or fuse in the box. To cut off the current entering your home, you don't have to haul on a switch handle or pull out a block. You just flip the main circuit breaker to *off*. To cut off power in a branch circuit, you do not have to unscrew a fuse. You just flip the smaller circuit breaker serving that branch circuit.

Circuit breakers have two disadvantages. First, they cost more than fuses of equivalent ratings. However, unlike fuses, circuit breakers are not ruined every time there is an over-

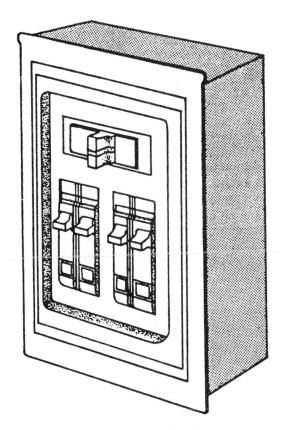

Circuit-breaker service entrance. Main breaker at top controls all current. The four other circuit breakers control four branch circuits.

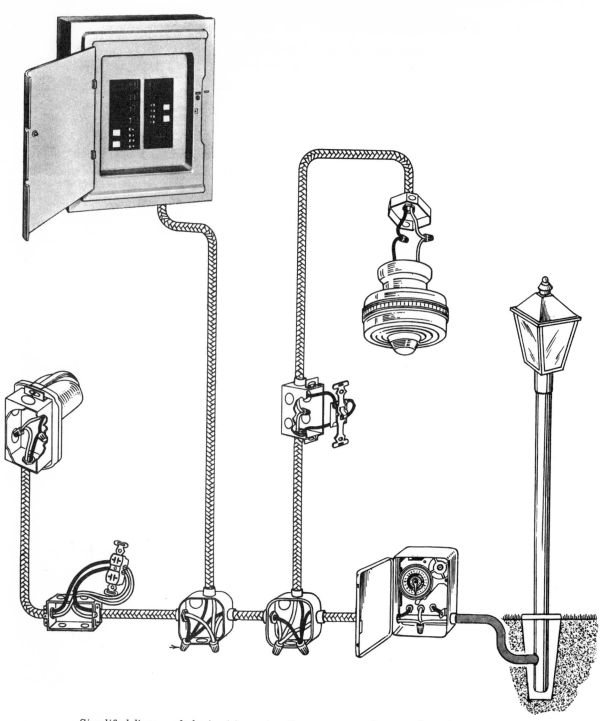

Simplified diagram of a basic wiring system. Current enters and returns through a service entrance. The current then runs through branch circuits — not unlike water through pipes. Typical components are shown. Two junction boxes (center bottom) are where wires meet and branch off. On one branch circuit (to left of junction box) there is a double outlet, and above it there is a wall-light fixture, in this case for a kitchen. Circuits from the other junction box lead to the switch and ceiling fixture (center) as well as to an automatic time switch and outdoor post lantern. The connecting cable contains two wires: a black or hot wire carrying current from the service panel, and a white ground or neutral wire that returns the current to the panel.

load or short circuit. They merely trip to the *trip* position. Therefore, less money is spent on replacing them.

The second disadvantage of circuit breakers is only a bit more serious. They are less helpful than fuses in diagnosing circuit defects (see Chapter 2).

Mapping Branch Circuits

Getting to know your fuse box or service entrance means getting to know which branch circuit is guarded by each fuse or circuit breaker.

A chart giving the desired information may be attached to the inside of the box cover. Electricians sometimes map out such a chart after they wire a house or an apartment. Or a chart may have been left by a previous owner or tenant.

A chart can prove helpful when trouble develops. So if you don't already have one, why not make one? Here's how to go about it:

1. On a piece of paper, show the location of every fuse or circuit breaker in the box.
2. For safety, put on thick rubber gloves. Turn the main switch or main circuit breaker to *off*.
3. Unscrew and remove one small round fuse — or flip one of the smaller circuit breakers to *off*.
4. Turn the main switch or main circuit breaker to *on*.
5. Next, walk through your home or get a helper to do so. Test every outlet by plugging a lamp into it. Test every wall and ceiling fixture by switching it on.
6. On your paper, note which outlets do not work and which fixtures do not light. Flag these to the location of the fuse you removed or the circuit breaker you flipped to *off*.
7. Return the main switch or main circuit breaker to *off*.
8. Replace the small round fuse or flip the smaller circuit breaker to *on*.
9. Repeat the procedure, checking out the rest of the fuses or circuit breakers in the box one at a time.

10. After your chart is complete, tape it to the inside of the box cover.

For Optimists and Others

If you are the kind of person who never anticipates trouble, or if you keep yourself busy with other things, or if you are just plain lazy — you may never bother to make up a chart. No matter. A chart is not vital; it's merely a convenience. If electrical malfunction does strike, there are other ways to identify the problem circuit and its fuse. But if the problem occurs during the night, you may find yourself poking your way through dark rooms, trying to find out which lights or outlets work and which

Never reach into a fuse box if wiring terminals are exposed. It's a good way to give yourself a shock you might not survive. Terminals are supposed to be completely covered by a plate — which sometimes is carelessly left off.

don't. Or, maybe you'll be in a hurry to know which appliances are plugged into which circuit because one of them keeps blowing its fuse to your television set — right in the middle of the bowl game. On such occasions, you'll regret not having a chart to consult.

Whether you are optimistic or pessimistic, lazy or not, **never forget that the only way to begin repair of a defective circuit is by switching off the current in the circuit.** Otherwise, you may electrocute yourself. At the very least, if the current is not switched off, your tools might cause a short circuit — showering sparks and possibly making the trouble worse by carbonizing connections. This warning applies to repairs involving any outlet, wall switch, light or ceiling fixture, cable connection — or anything else permanently wired into your household electrical system.

Of course, you need not turn off any house current to repair a plug-in lamp or a plug-in appliance. However, **such devices must be unplugged before you begin to repair them.** Once its plug is removed from the outlet, of course, the appliance has no kick. You can work on it safely.

Alternative Cut-Off

Should any electrical repair or replacement become necessary, your chart will tell you at a glance which circuit to kill — and which fuse or circuit breaker can do the killing. If you don't have a chart, you can trace out your circuits then and there, using the method described for making the chart (see page 11). Or, chart or no chart, circuit knowledge or no knowledge, you can proceed safely by pulling the main block and throwing the main switch to *off* or by flipping the main circuit breaker to *off*. If your service entrance contains more than one main block or main circuit breaker, pull them all or flip them all to *off*. To be extra safe, do the same thing to any blocks or circuit breakers marked *range*.

Now every circuit in your home is dead and you are safer than safe.

If you live in an apartment with a separate panel that does not have a main switch or main circuit breaker, simply open the cabinet door

and remove every fuse you see or flip every circuit breaker to *off*. This will deaden the branch lines as effectively as throwing a main switch. You can then tackle any repair safely.

Eliminating all the current flow within your home, while useful in emergencies, has obvious drawbacks. While you are repairing an outlet or replacing a broken wall switch, no one in the house can use a waffle iron, a hair dryer, or any other appliance. If you make your repairs at night, you will have to work by flashlight while the rest of your family suffers along in darkness.

That's why making a circuit chart, if you don't have one already, remains the sensible thing to do. It gives you speed and flexibility in an emergency.

Diagnosing Troubles

Apart from the failure of plug-in lamps and appliances, most electrical difficulties in a home arise from either (1) a short circuit or (2) an overloaded circuit. Rarely, the difficulty is traced to a so-called *open circuit* — a break in the wiring or the failure of contacts to meet.

As previously mentioned, shorts and overloads generate heat that blows a fuse or trips a circuit breaker. When this happens, no light fixture or outlet on the affected circuit will work. Find a pair of thick rubber gloves and a flashlight. Then go to your service entrance, but do not touch any of the fuses or circuit breakers yet. If your service entrance contains fuses, follow the procedures given below. If your service entrance contains circuit breakers, follow the procedures listed under Circuit-Breaker Method (page 16).

Fuses Tell a Tale

If your service entrance contains common screw-in fuses, the blown fuses often give you information. Each fuse has a transparent face. Look through it, using the flashlight if necessary.

Does the face of the fuse appear darkened, powdery, and no longer transparent? Then most likely the fuse blew because of a short in the branch circuit it protects. Does the face of

Left: *fuse blown by short circuit*. Right: *fuse blown by overload*.

the fuse appear clear and can you see that the tiny strip of metal beneath the face has parted? Then most likely the fuse blew because the branch circuit it protects was overloaded.

Short Circuit or Overload?

Did anyone turn on a lamp or appliance just before the fuse blew? That can give you a clue. Maybe the lamp or appliance was faulty and shorted the circuit. Or, maybe the lamp or appliance drew enough additional current to overload the circuit, presuming that other appliances were already plugged into it and operating.

Go back and unplug all lamps and appliances from outlets on the defective circuit. How do you know which outlets those are? From your

chart. If you have no chart, the safest bet is to unplug every lamp and appliance in the house. Any wall or ceiling fixtures that refuse to light are on the affected circuit, too. Turn their switches to *off*.

Return to the service entrance. After taking proper precautions to avoid shock, unscrew the dead fuse. At this point, if you want to, you can screw a 75- or 100-watt bulb into the fuse socket. Does the bulb glow brightly? That indicates a short in the house wiring. Does the bulb remain dark? Suspect that an overload, rather than a short, blew the fuse — unless one of the appliances or lamps contains a short. You will have to test each, as described below.

But you can skip the bulb-method. In any case, a standard bulb cannot be screwed into a fuse socket adapted for a special-base type-S

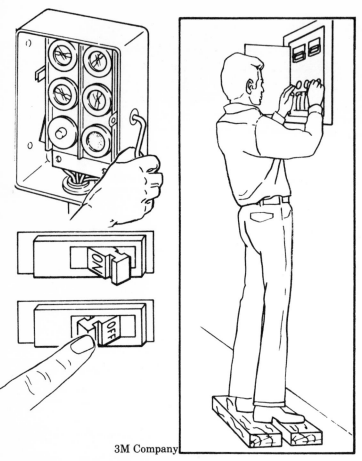

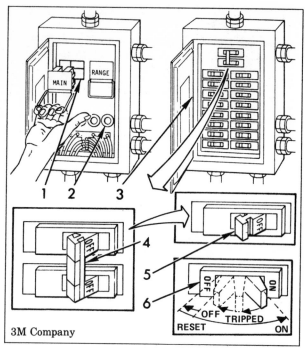

3M Company

The main switch may be a block that pulls out to shut off the house current (1). A blown fuse (2) will kill the current only in one branch circuit. A circuit-breaker service entrance (3) always has a larger or a double breaker (4) for a main switch. A tripped breaker is equivalent to a blown fuse. In this installation, breakers (4, 5) reset by flipping them first to off then to on (6). This is equivalent to replacing a blown fuse with a good fuse.

3M Company

Before changing a fuse pull the main switch or the main circuit-breaker to off. Don't stand on wet spot. Don't touch box or fuses if they are wet.

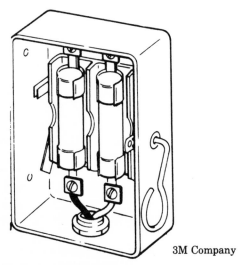

3M Company

In some homes, the main switch is located outside fuse box. Know the location of your switch.

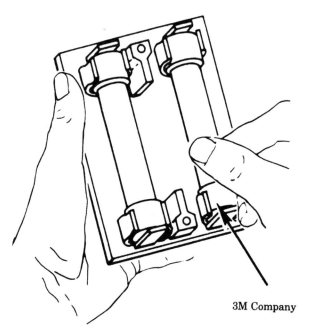

3M Company

Block contains cartridge fuse or fuses.

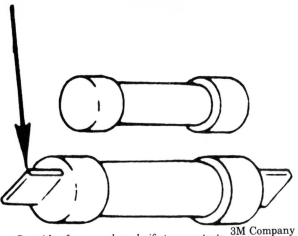

Cartridge fuse may have knife-type contacts   *3M Company*

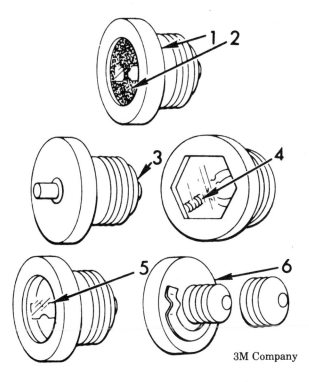

3M Company

Standard fuse (1) has an Edison base; that is, it screws into the same socket as a light bulb does. If blown by a short, its face will be darkened (2). This type of fuse (3) operates like a circuit breaker and can be installed in any fuse box. Spring-loaded fuse (4) has time-delay feature, which tells whether circuit has shorted or has been overloaded. When a standard fuse overloads, the metal strip parts at the weakest point (5). A type-S fuse (6) comes with a special base and adapter. The adapter locks into the fuse socket, preventing insertion of a fuse with the wrong base — and therefore of wrong amperage rating.

fuse. A more direct method, after unplugging lamps and appliances, turning off light switches and removing blown fuse, consists of:

1. Replace the blown fuse with a fresh one of the same amperage. If it blows immediately, you know the house wiring is shorted, probably at some outlet.

2. If the fresh fuse does not blow, ask someone to switch on, then switch off, each wall and ceiling fixtures. If switching on a fixture blows the fresh fuse, there is a short in that fixture.

3. If Step 2 does not blow the fresh fuse, you or your helper should plug in and then unplug each lamp and appliance in turn. Make sure each is switched on. If plugging in a lamp or appliance blows the fuse, there is a short in that lamp or appliance — but no short in the house wiring (for appliance repairs, see Chapter 7).

4. If Step 3 above does not blow the fresh fuse, you can be pretty sure that an overload blew the original fuse. If you wish, you can verify this by plugging in all of the lamps and appliances one by one, letting them stay plugged in. At some point, probably the fresh fuse will blow. This confirms an overload.

How to Change a Fuse

1. Make sure you are not standing on a wet spot. Wear thick rubber gloves, if possible.

2. Turn the main switch to *off,* or pull out the main block.

3. Remove the fuse by turning it counterclockwise.

4. Screw in the new fuse by turning it clockwise.

5. Turn the main switch to *on,* or push the main block back into place.

(Apartment residents may not have a main switch. It is perfectly safe for them to change fuses provided no bare wires, terminals, or connections are visible within the fuse box when the door is open — and provided they do not touch the metal of the fuse socket. Here rubber gloves are an all-important precaution.)

Circuit-Breaker Method

If your service entrance is fitted with circuit breakers rather than fuses, follow the same procedure as that given above for fused entrances. With a difference, of course. You cannot remove or replace a fuse because there aren't any.

So where the directions tell you to "unscrew" or "remove" a fuse, simply flick your circuit breaker to its *off* position. Where the directions say to "screw in" or "replace" a fuse, reset the circuit breaker to its *on* position. Where the directions mention a fuse blowing or not blowing, that means to you a circuit breaker tripping to *off* or remaining at *on*.

Note that the National Electric Code requires different circuits for outlets and light fixtures in rooms where appliance current drain is likely to be high. Kitchens, laundry rooms, and utility rooms fall into this class. So if a circuit serving your kitchen, for example, blows a fuse, you know that probably either outlets or light fixtures are on that circuit, not both.

Open Circuit Indication

If a light or outlet fails, but you find that none of the fuses have blown, or no circuit breaker has tripped, suspect an open circuit. There may be a break in a wire. Or, contact may have been lost between a wire and a terminal, one terminal and another, or the prongs of a plug and the receptacle slots of an outlet. In any case, current will not flow because the path through the circuit is broken. This is similar to what happens when you put a switch in the *off* position. The current is cut off by "opening" or "breaking" the circuit.

Most breaks lie in the particular unit that does not function. If more than one light or outlet on the same circuit will not function, you may have a break in the house wiring (see page 27).

Practical Considerations

Remember that it may take a while for an overloaded circuit to build up enough heat to melt the special metal within the fuse or to trip a circuit breaker. So when switching something on to test for overload, allow plenty of time before switching on the next thing. An interval as long as 30 seconds may be required.

You can save yourself annoyance by investing in a simple device called a *high-voltage circuit tester* (also called a *line tester* or a *continuity tester*). Quite inexpensive, this gadget consists of a tiny neon bulb and a pair of insulated probes. Remember when in making a circuit chart or in figuring out which circuit was alive or dead after a fuse blew you had to walk around plugging in lamps and appliances? A high-voltage circuit tester will tell you immediately whether an outlet is alive or dead. It also can tell you if an outlet is properly grounded, which wires are hot and which are neutral, and whether units are shorted or open.

Since these testers draw their power from the house current, they are designed to operate at 110 volts or more. They should not be confused with low-voltage continuity testers, which are similar in appearance but powered by flashlight batteries. Low-voltage testers are also a great convenience — particularly in

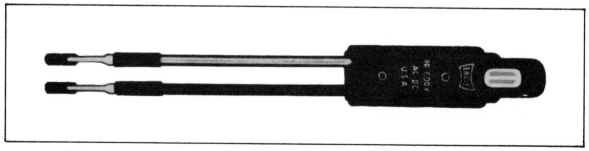

High-voltage circuit tester. The bulb will not light if circuit is broken. It can be used for automobile work, too.

chasing down shorts and breaks in lamps, appliances, and their line cords. Both types of testers are inexpensive and can be bought at any hardware or electrical supply store.

You probably have realized by now that, with or without tools like testers, diagnosis of electrical malfunction is chiefly a matter of using your head, since trial-and-error is often the best procedure.

Safety is also largely a matter of using your head. Wear thick rubber gloves, if possible. Use pliers and screwdrivers with insulated handles. Never touch anything that might have live electricity running through it, particularly bare wires and bare terminals. Avoid wet spots. Above all, make sure the fuse controlling a circuit is removed before you work on any wire, outlet, switch, or light in that circuit; better still, keep the main switch or main circuit breaker at the *off* position.

With those considerations in mind, let's move on to some actual repairs.

2
Correcting Shorts and Open Circuits

Suppose that, by following the procedure outlined in Chapter 1, you have established that a short exists in a branch circuit. Further, suppose that you have pinpointed the fault to a particular wall outlet or wall switch (when you plug something into the outlet or when you throw the wall switch to *on,* the circuit's fuse blows or its circuit breaker trips).

Almost always, the surest means of repairing an inoperative outlet receptacle or switch is to replace it. This is not difficult if you follow the simple step-by-step procedures listed below.

1. Eliminate any chance of being shocked while you work. Cut off all current by throwing your main switch or main circuit breaker to *off.* Remove the fuse protecting the circuit or move its circuit breaker to the *off* position. If it is necessary to keep other circuits in your home operating, you now can return your main switch or main circuit breaker to *on.*
2. Remove the faceplate covering the outlet or switch.
3. Remove the holding screws (the screws that hold the unit to the metal box behind it).
4. Loosen the terminal screws and detach all wires, noting which wire goes to which terminal. Discard the old unit.
5. Sandpaper the bare wire ends clean of dirt and corrosion.
6. Attach the wires to your new unit exactly as they were attached to the old — that is, black or red wires (hot) go to brass-colored terminals; white wires (neutral) go to silver-colored terminals.
7. Push the wires and unit back into the box; reinstall the holding screws and the faceplate.

Ground Arrangements

Outlet boxes, switch boxes, and junction boxes should be grounded independently. If your home is wired with metal-covered two-wire cable (known as BX), the metal extending from box to box finally winds up at your service entrance. There the metal is connected to a rod sunk into the ground or to cold-water pipes that run underground. If your home is wired with cable run-

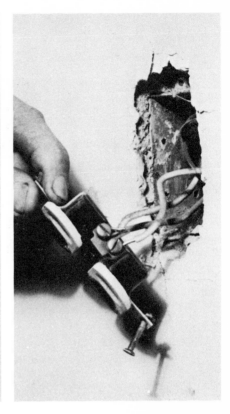

Shut off the current at main switch or circuit breaker. Remove faceplate. Loosen the two holding screws at the top and bottom of the outlet. This duplex unit is cracked and chipped — a sure source of trouble.

Pull outlet from the wall box.

Disconnect outlet by loosening the terminal screws.

Hook up new outlet like the old one: black wire to brass terminal, white wire to silver-colored terminal. A third wire, colored green, grounds the outlet to the box (not present in all outlets).

Install the outlet by screwing holding screws through rabbit ears and into box. Put back the faceplate and the job is complete.

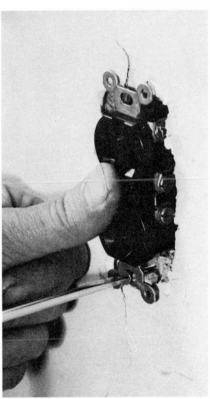

Outlet faceplate (also called wallplate) goes on or comes off with the turn of a single screw. There are 2 screws on a switch faceplate.

A switch is removed or installed in the same manner as an outlet.

ning through metal conduit, the conduit extending from box to box is grounded similarly.

But your home may be wired with non-metallic cable. If so, there should be a third wire in the cable, usually bare copper, that serves to ground the box through the service entrance. When replacing a unit, make sure that any bare wire coming into the box is firmly fastened to it by clip or screw.

For three-prong grounding plugs, this duplex outlet has a self-grounding feature, eliminating need to hook up grounding wire. Corrosion-resistant long screws, colored green and fixed firmly by spring action, ground the unit to the box.

Today, many local electrical codes require that the unit itself, in addition to its box, be grounded. Many makes of units with ground connections are available from hardware and electrical supply stores. It's an excellent idea, when replacing a unit, to install one of the new grounding types. All you have to do is run a short length of wire from the ground connection on the unit (usually the darkest screw) to a clip or screw that fastens the wire to the box. If you are replacing an outlet that has a three-prong "grounding" plug, the grounding connection is vitally important. Note that most boxes contain a hole into which a threaded or sheet-metal screw can be inserted to grip firmly the bared end of the green grounding wire. If there is no hole, you can buy special clips for the purpose. One type of three-hole outlet is automatically grounded to the box by its green-colored holding screws, eliminating the need for a wire connection.

Understanding Grounding

The term "grounding" applies to a safety feature that should be built into every circuit carrying enough electricity to be dangerous. Grounding prevents accidental electric shock — and, incidentally, guards against damage by lightning.

All electricity, including lightning, follows the shortest and least resistant path to the ground. So if you should accidentally touch the bare part of a hot wire, any current it carries might jump right through you to the ground — inflicting shock. Shock is always unpleasant, and it can also be dangerous or even fatal.

Sometimes frayed insulation or movement of bared ends causes a hot wire to touch a metal surface. Unsuspectingly, you also touch the same metal surface. Again, the result may be shock. To avoid such risks, all metal surfaces around a hot wire should be grounded. That is, they should have an independent route to the ground by means of wire, conduit, or some other conductor. Then, if a hot wire happens to touch a metal surface, the current will speed directly to the ground. Should you touch that same surface, you will feel no shock. The current bypasses you because the path through you to the ground is longer — and electrically more resistant — than the path through the conductor (unless, of course, you are standing on a wet spot or lying in your bath; water is a pretty fair conductor).

As for lightning — a bolt of raw, enormously powerful electricity — picture it crashing down on a home. If it encounters grounded boxes and grounded cable, all or at least some of its voltage may run off harmlessly into the soil, preventing extensive damage.

After a replacement device is installed, you can check that it is properly grounded with a high-voltage continuity tester. Instructions for checking a ground may come with the tester when you buy it. If, for any reason, you are not certain that the units and boxes in your home are properly grounded, consult an electrician. Grounding is essential for preventing accidental shocks, particularly when using appliances or electric tools.

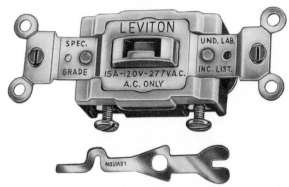

To keep potentially dangerous shop tools or other apparatus from being activated accidentally, locking switches are available that can be turned on only with a key. This tamperproof unit is guaranteed for twenty-five years.

Alternative Units

You can cut off the current in a circuit and then substitute one type of unit (or device) for another at any time. Simply follow the instructions already given for replacing an old unit with the same type of new unit. However, a particularly good time to change a unit occurs when you find that you have a short or some other defect. You must take off the faceplate and replace the unit anyway. This is also a good time to replace an old unit with a different type.

All outlet and switch units come in standard sizes, so virtually any one can replace any other. And there is a broad range of single and combination units on the market today, allowing a great range of choice for the do-it-yourself householder.

For example, you might want to substitute a quiet switch or a silent (mercury) switch for your old one. Or, you might want to replace a single outlet with a double outlet. There are also many other possibilities for you to consider. These include: nightlight plus outlet, switch plus outlet, pilot light plus switch, dimmer switch, rocker switch, switch or outlet with a grounding terminal, etc.

Amperage Ratings: When buying a new outlet or switch, make sure it has an amperage rating no greater than that of the unit being replaced.

This outlet has built-in protection against any fault in grounding that might develop in a branch circuit. Unit also tests for such faults. Decorative faceplate is available for indoor use.

For outdoor use, as in guarding a swimming pool, this Pass & Seymour duplex outlet is the last word in reliability. Spring-loaded metal hood prevents water-seep and weathering.

Replacing standard toggle wall switch with a modern dimmer switch. The only tool necessary is a screwdriver.

Dimmer installed with faceplate. Unit allows adjustment of room illumination to any level from off to full bright.

Most devices have their ampere capacity plainly stamped on their faces. Inserting a 15-amp unit in a 20-amp branch circuit would be folly, since the 15-amp unit would overheat. If you don't know the ampere capacity of the circuit you are working on, remember that most home circuits are rated at 15 amperes, that kitchen and utility circuits are rated at 20 amperes, and that special circuits (for serving motors, shop tools, or heavy appliances) may be rated at 30 amperes.

For safety, motors, shop tools, and heavy appliances (whether connected to a 20-amp or a 30-amp circuit) should always have a grounding line cord ending in a grounding three-prong plug. However, equipment having a three-prong plug does not necessarily have to be plugged into a three-hole outlet. With the aid of an *adapter,* the three prongs can be used with a two-hole outlet. An adapter has two prongs that fit into a common two-hole outlet, but it also contains three holes to accommodate the grounding plug. The connection is grounded by means of a green wire extending from the adapter to the outlet. Loosen the screw holding the faceplate, slip the bared end of the green wire under the screw, then tighten it. Provided the outlet box is grounded (as it should be in a properly wired home), this grounds the equipment.

Interchange Devices: Switches, outlets, built-in pilot lights or nightlites, or any combination of these, usually are built as units. They may consist of many pieces, but are fabricated into what is, in effect, one piece. Undoubtedly, a single unit is easiest for a do-it-yourselfer to work with, but such units may present problems. For instance, if you want to install two outlets and a switch in a box previously accommodating only two outlets, the box opening may not be large enough to hold all three devices. Or, perhaps you already know that you will want to change the two outlets and a switch to one outlet and two switches later.

The solution to this and other such problems is the so-called *interchange device.* Although interchange devices come in all the varieties found among regular combination units, each device comprises its own unit — and one is exactly the same size as another. So any inter-

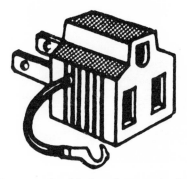

Adapter for converting 2-hole outlet to receive 3-prong grounding plug. The green wire extending from adapter body is fastened under the screw of faceplate.

change device can be substituted easily for any other. You merely snap them in or out of a special strap designed for these devices; the strap fastens to the box in your wall. Furthermore, interchanges are built compactly, so that, for example, three can be installed where only a single regular unit would fit before.

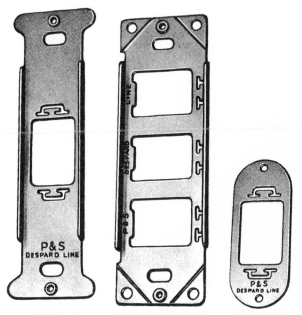

Special straps fasten to box and hold interchange units in place.

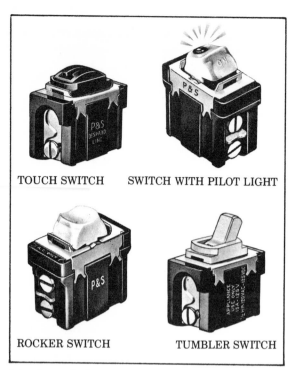

TOUCH SWITCH   SWITCH WITH PILOT LIGHT

ROCKER SWITCH   TUMBLER SWITCH

Interchange devices are compact, convenient. Any one can replace any other.

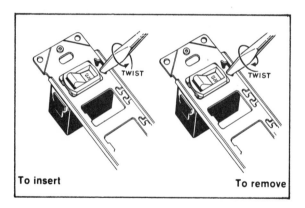

To insert   To remove

Faceplate Change: Faceplates, commonly called wallplates, may be made of metal, plastic, or some other material. They come in an almost infinite variety of styles, both plain and elaborate, and many are quite pleasing and decorative. They are available in standard sizes and oversizes — some with specially deep flanges — and with an endless assortment of openings.

If you are replacing a unit with one of the same type or one that is similar (for example, replacing an ordinary switch with a mercury

Switch and grounding outlet

Two switches

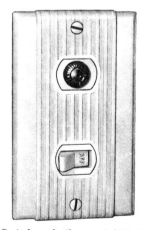

Switch and pilot or nightlight

These are "Despard" faceplates for interchange combinations.

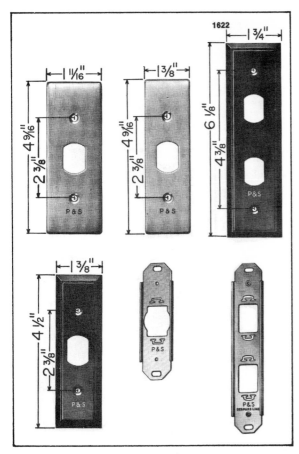

Faceplates for interchanges come in special narrow sizes for use where space is limited.

switch), there is no need to change the faceplate. But if you replace a unit with one of a different kind (for example, replacing a double outlet with a combination of outlets and a switch), then you will have to install a new faceplate with the appropriate openings.

Sometimes boxes are not sunk deep enough into the wall, so that when a faceplate is screwed into place it fails to make contact with the wall surface. A gap between the edges of the wall and the faceplate results. This condition can be corrected by installing a faceplate with extra-deep flanges that will reach out and touch the wall. Otherwise, you will have a gap that is not only unsightly but also hazardous. A gap would permit dust and foreign material to work their way into the unit and box. And if a child should happen to poke a hairpin or some other metal object into the gap, a short and, therefore, a shock might result.

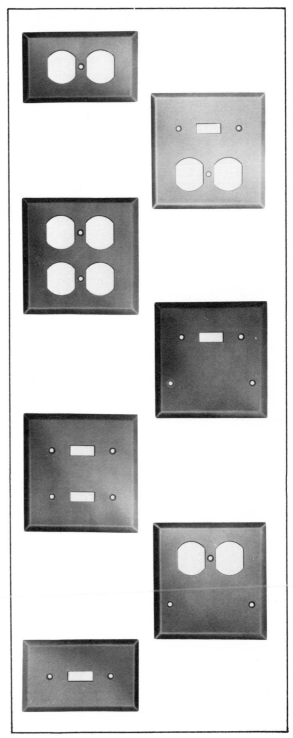

Extra-deep faceplates come in nearly as many styles and hole-combinations as standard faceplates do. They solve the problem of covering boxes in shallow walls and thin partitions, where the box may lie flush with the surface or even protrude.

Shop around before making substitutions that involve interchange devices. Some stores offer the necessary faceplate packed with the corresponding mounting strap. This saves you trouble and makes sure you wind up with a correct fit.

Oversized faceplates are also sold in stores. These come in handy when you want to cover mars or cracks in the wall surfaces adjoining box openings.

New Wiring: Substituting one kind of wiring device for another may require changing the wiring. Usually the required change is slight. Sometimes a wiring diagram comes with the new unit when you buy it. If not, the salesman in your electrical supply store can tell you what to do.

Make certain, though, that the brass-colored terminals on your outlet units connect only with black wires and that the silver-colored terminals connect only with white wires.

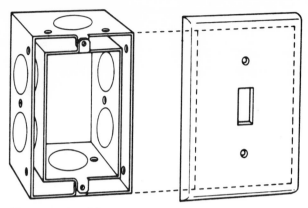

If a conventional faceplate cannot cover a new box, or if the wall is chipped around the box, you can hide all with an oversized faceplate.

Grounding outlets, rocker switches, and faceplates can be sleek and modern. This is the Decora line.

Switches should be connected to black or red wires only. Switch equipment should never be inserted into the neutral or ground side of a circuit, the wires of which are white.

The Open Circuit

Since an open circuit is caused by a break of electrical contact somewhere in the wiring or terminals, it is often called a "broken" circuit. As mentioned previously, when you have an open circuit, your outlet, switch, or other electrical apparatus will not work, but this malfunction, unlike that of a short circuit or an overload, does not blow a fuse.

Standard faceplate   *Oversized faceplate*

These are Midway faceplates of the Snapit line.

It is rare for an open circuit to occur in the house wiring (the lengths of cable running for considerable distances through walls or buried in floors). If this should occur, there is little the amateur can do — it is best to call in a licensed electrician.

However, there is one type of open circuit that could occur in the house wiring that you may wish to fix yourself. The cables used to wire a house usually terminate in a square, rectangular, octagonal, or round junction box constructed of steel or some similar metal. In the junction box, the wires of one cable are connected to the wires of other cables by means of solderless connectors. Each connector consists of a cone-shaped plastic cap with a coiled copper insert. The two or more wires to be fastened are twisted together and then inserted into the cap. A firm and locked connection is made by turning the connector clockwise. The connection is insulated by the cap.

Although it rarely happens, a solderless connector can work itself loose because of exces-

Junction box exposed on basement ceiling. Note cable connectors locking BX at knockouts (see Chapter 4).

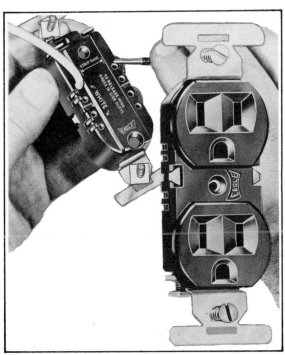

You can buy outlets and switches that make wiring quicker by means of pressure terminals rather than screw terminals. Some units have both kinds built into them and give you a choice of which to use. To connect wire in the pressure type, simply poke the bare end into the hole. To disconnect, insert screwdriver into the side slot.

Junction box opened to reveal white wires, red wires, black wires, and solderless connectors. All bare wire is covered.

sive vibration or for some other reason. The result can be an open circuit (or even a short) within the box. If boxes are exposed, as in basements and attics, you can unscrew their covers and inspect the connections to see that they are tight. **It is unsafe to remove any junction-box cover, however, unless you first turn off your main switch or main circuit breaker.**

Equipment Breakdown

If the break of contact causing the open circuit does not lie in the house wiring, you can usually locate it by inspecting outlets and switches.

A short generally is revealed by a heavily carbonized spot, and perhaps, by burned insulation. Also, a short gives off a characteristic acrid smell. An open circuit, on the other hand, may be caused by corrosion around the unit's screw terminals or by wires that have worked loose from the terminals. The latter, an open circuit caused by loose wires, can be corrected by reconnecting the wires and turning the terminal screws tightly. If an open circuit is caused by corrosion, it is best to install a new unit. When installing, sandpaper until they are clean and bright.

Simple mechanical failure can also cause an open circuit. For instance, the unit may have become cracked, chipped or badly worn. In time, the receptacle contacts of an outlet may lose their grip. Paint may have worked its way into them during a decorating job, insulating them from contact with plug prongs. The lever of a toggle switch may freeze. In all such instances, it is best to replace the unit.

Open as well as short circuits can cripple ceiling lights, wall lights, lamps and, of course, any appliance. What to do when such apparatus fails is discussed in later chapters.

Working on a lamp or appliance is always perfectly safe because once it is unplugged from its outlet no live electricity can get into it. There is no chance of shock. *But never work on an outlet, switch, junction box, light fixture, or other apparatus permanently linked into an electric circuit in your home without totally squelching the current.*

In other words, first turn your main switch to *off,* pull your main block right out of the service entrance, or flip the main circuit breaker to *off.* Then remove the circuit's fuse or flip its circuit breaker to *off.* Be sure you remove the right fuse or flip the right circuit breaker. Once that is done, it is safe for you to work even with your main current restored. If in doubt, or to be double-safe, do not restore the main current.

Fuse Defects

Sometimes an open circuit is caused by a defect in the fuse or the fuse socket. You should suspect such a defect when a branch circuit does not work but the window of its fuse remains clear and the strip of fuse metal beneath it appears intact.

After making certain that the main switch is *off,* unscrew the suspect fuse. Look into the fuse socket. At its bottom you will see a screw. If this screw is loose, it may be causing the open circuit. Turn it down tight.

Look the fuse over. If the contact on its bottom is corroded or pitted, this too is evidence of a faltering connection. Replace the defective fuse with a new one of exactly the same amperage.

Never insert a penny or tinfoil in the fuse socket. This age-old trick to keep the circuit going without a fuse is dangerous. Unfused circuits can burn out wiring, destroy equipment, and cause fires.

3
How to Eliminate Overloads

Each branch circuit in your home has a certain fixed electrical capacity that can be stated in watts. The number of watts a circuit can carry is found by multiplying its amperage rating by the voltage of the current supplied. Therefore, a 15-amp circuit can handle 1725 watts (15 × 115), a 20-amp circuit can handle 2300 watts, and a 30-amp circuit can handle 3450 watts. If the electrical capacity of any circuit is exceeded, an overload results — and a fuse blows.

In practical terms, the difficulty arises this way. Suppose that you plug a broiler using 1500 watts, a coffee perker using 500 watts, and a blender using 300 watts into a 20-amp kitchen circuit. When you switch on these three appliances together, they draw 2300 watts (the full capacity of the circuit). If you now plug a 1000-watt frying pan into the circuit and switch it on, an overload occurs and the fuse blows.

Similarly, you can use a 100-watt ceiling light, a 75-watt wall light, a 1200-watt room heater, and a 300-watt television set (a total of 1675 watts) in your bedroom without blowing the 15-amp fuse. But if you then switch on a vacuum cleaner pulling 500 watts, the total draw is 2225 watts, exceeding the circuit's 1725-watt capacity. First the lights dim and then the fuse blows.

Solutions Are Few

There are only three solutions to overload problems. The first is obvious: simply do not plug so many devices into any circuit that they exceed the capacity of the circuit. Even though this may prevent you and your family from using modern electrical appliances and lamps whenever you wish, it will solve the problem. The table below will give you a general indication of which appliances — and how many — you can plug into a circuit without exceeding its capacity.

Watt Consumption of Typical Appliances

Appliance	Watt consumption
Air conditioner, small room	750
Air conditioner, medium room	900
Air conditioner, large room	1300
Air conditioner, central (230 volts)	5000
Blanket	150
Blender	250–450
Broiler	1500
Can opener/knife sharpener	100
Ceiling light, per bulb	40–150
Clothes iron	1000
Clothes iron and presser	1700–3000
Clothes dryer, all electric (115 volts)	1400
Clothes dryer, all electric (230 volts)	4500
Clothes washer, automatic console	850
Coffee percolator	500–750
Deep fryer	1350
Dehumidifier	400–600
Desk light	75–100
Dishwasher, heating type	1800
Drill, portable	200–400
Drill press	500
Fan, kitchen ventilator	70
Fan, large floorstand-type	600
Fan, table type	75
Floor polisher	300
Fluorescent, per tube	20–40
Freezer	350
Frying pan	1000
Garbage disposal	900
Hair dryer	250–600
Heater, portable space	1325
Heater, space	1700
Heating pad	60
Hot water heater (230 volts)	2500
Lamp, floor	75–300
Lamp, table	60–150
Microwave oven	650
Mixer	150
Nightlight	7½
Oil furnace (230 volts)	1200
Radio, table, with tubes	50
Radio, table, solid state	10
Range (230 volts)	8000–16,000
Refrigerator	250–300
Rotisserie	1400
Saw, radial	1500
Saw, table	600
Shaver	10
Soldering iron	150
Television, black-and-white, 18-inch	300
Television, color, 18-inch	350
Toaster	1100
Ultraviolet lamp	100–400
Vacuum cleaner	400
Wall light, per bulb	40–100
Waffle iron	1100

For more specific information, consult the wattage rating stated on the nameplate, bottom plate, or elsewhere on the housing of a particular appliance. Motors are rated by horsepower (HP); one HP equals approximately 750 watts.

The second way to solve a chronic overload problem is by tapping additional outlets into another branch circuit that is not being used to its full capacity.

The third way is to install one or more additional branch circuits as well as the outlets they will feed.

Do It Yourself — Or Call a Pro?

Before you decide to add outlets or branch circuits to cure overload, there are important things to consider. The work can be difficult. It requires tools, patience, and a certain amount of mechanical skill. Far more important, such work is subject to the regulations of your local power company, your local electrical code (whether municipal or state), and the National Electric Code. The latter consists of rules approved by the National Board of Fire Underwriters to assure safe wiring. All cables, wires, outlets, switches, and lighting fixtures should carry a stamp or tag indicating approval of Underwriters' Laboratories (UL). Just *using* UL-approved materials is not enough, however; follow installation rules precisely.

Your local power or utility company, as well as your local building inspector, can supply you with copies of the applicable regulations. When it comes to cutting into and going behind walls, ceilings, and floors or changing or adding to the basic circuitry, you probably will find that your locality has strict regulations. Most often you will find that the completed work must be officially inspected and approved before it is used. In most places these regulations do not apply to repairing or replacing a switch, outlet, lighting fixture, or anything else connected to — yet not integral to — the basic electrical circuitry of your home.

So even if you have a constant overload in your home, consider carefully whether or not you want to fix the problem yourself or turn it over to a competent and reliable electrician.

Hooking into a Branch Circuit

If you decide to solve the overload problem yourself by linking additional outlets to a circuit, you must begin by selecting a branch circuit that is not being used to its optimum load. Linking the additional outlets to this circuit involves three main steps: (1) an opening must be cut in the wall to accommodate each outlet box, (2) wiring cable must be installed for connecting the outlet to the circuit, (3) the wiring connections must be made.

While you want the new outlets to be placed as conveniently as possible, the sites you can use are limited. The outlets can be positioned only where it is possible for you to run cable from them to the branch circuit. You will find it most practical to hook into the circuit at an old outlet already in the wall, at a light fixture, or at an exposed junction box.

Common sense must be your guide to finding the easiest path for the connecting cable — the path that will cause the least havoc to your home structure. Because house construction varies, no exact rules can be set forth. But by moving the intended location of your outlet to another point, you may be able to avoid a difficult drilling job or fishing operation. (Fishing is the term for threading cable through walls, floors, or ceilings — a procedure that can fray even the stoutest nerves.)

Placing the Opening

When you buy a new outlet and outlet box, you may also get a template for the outlet box. At the spot where you wish to install the outlet, press the template to the wall and pencil its outline. If you have no template, press the box itself to the wall and pencil its outline.

But before deciding on the exact spot for your

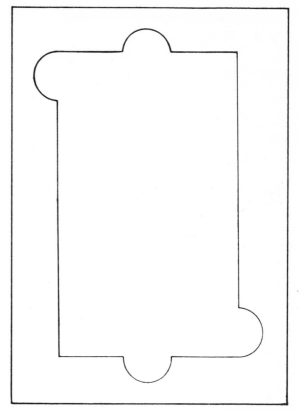

Full-sized template for standard outlet and switch boxes.

new outlet, try to find the stud behind the wall nearest your chosen spot. Tap lightly against the wall; a solid rather than a hollow sound tells that you have found a stud. Or, use a stud-finder, which can be bought in a hardware store, to find the stud. Next, move your approximate box location far enough left or right to clear the stud. Then drill a small hole through the wall and probe with a stiff wire to make sure that the box will clear the stud. (In a pinch, you can always locate studs by pulling off your baseboard and looking behind it — or by drilling small, inconspicuous holes at 2-inch intervals along the top of the baseboard.)

When penciling the outline of the box, bear in mind that you will have to fasten the box in place permanently. In dry-wall construction, the box can be affixed to the stud itself (this is the strongest anchoring) or to the dry-wall. In plaster-and-lath construction, the box can be fastened to the lath. If you intend to use the stud as an anchor, cut the opening so that the box will sit right next to the stud. If the box

You can buy boxes fitted with a bracket for fastening to stud.

will be anchored to the dry-wall or the lath, cut the opening a few inches to the left or right of the stud.

Dry-Wall Installation: You must buy an outlet box suitable for the kind of installation you contemplate. If your home has dry-wall construction, and you wish to fasten the box to the stud, buy a box with a bracket extending from the side. On the other hand, if you wish to fasten the box to the dry-wall, buy a box with adjustable mounting "ears" at the top and bottom of the box.

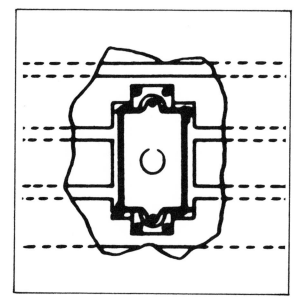

Plaster laths cut to hold box.

Plaster Installation: Since mounting ears can be adjusted and even inverted, they can anchor a box not only to dry-wall but also to plaster. A much stronger way to anchor in plaster, however, is to screw the box to the lathing behind the plaster. So after you have decided where to place the box, dig out a small section of plaster until you locate the laths. If the laths are wooden, cut a section just wide enough to accommodate the box out of one lath. In the lath above, cut a notch to accommodate the top of the box, and in the lath below, cut a notch to accommodate the bottom (see illustration).

Some construction seats the plaster on gypsum laths. In such cases, use either a box with screw-type lugs or a plain box plus a plate with tabs. The tabs lock the plate to the box and anchor it to the gypsum.

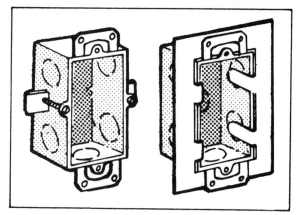

Gypsum lath requires box with side lugs for anchoring — or a plate with tabs that bend into box.

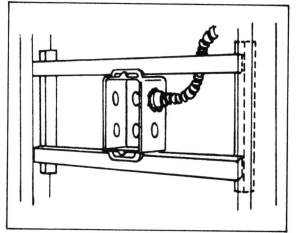

One or two wooden cross-members can also anchor box to stud. This method requires cutting and patching wall.

How to Install a Box Opening in Plaster or Dry-Wall

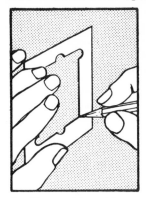

1. Trace outline with template or by profiling box.

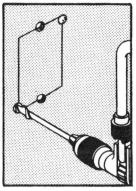

2. Bore ¾-inch holes top and bottom and on diagonal.

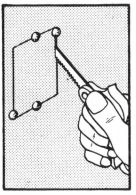

3. Start keyhole saw in holes and cut along outline.

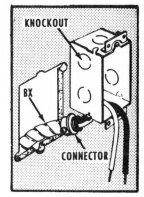

4. Bring cable into box; connect before fixing box in place.

(If there is no other way to solve your anchoring problem: (1) cut away a horizontal section of plaster between two studs, (2) angle-nail a horizontal length of 2 × 4 to the two studs, (3) complete wiring and screw box to the 2 × 4, (4) fix horizontal cut by patching the plaster.)

Cutting the Opening

After outlining the exact location of the new box on the wall, drill a ¾-inch hole at the top of the box and a similar hole at the bottom. These holes allow entry of the holding screws which bind the outlet unit in place. Drill two diagonally placed holes to give the saw blade a starting place, then, using a hacksaw blade or small keyhole saw, cut along the outline.

Saw slowly and carefully. Plaster tends to crack unless you hold a board against it with one hand while you stroke the saw blade toward — not away from — yourself.

Laying the Cable

The most difficult part of connecting a new outlet to a branch circuit is stretching the cable between them. You may have to run the cable through walls, over ceilings, and/or under floors.

The most practical place to hook into the branch circuit is at an existing outlet, light fixture, or exposed junction box. If you place your new outlet directly above an existing out-let on the branch circuit, stretching the cable is fairly easy. Feed the cable into the new opening, allow the cable to drop down between studs, and grab it when it reaches the old outlet.

WARNING: Never attempt cable work unless all of the electricity in your home is shut off. Pull the main switch to *off* and pull out the main block or set the main circuit breaker at *off*. Otherwise, accidental contact may be made with a live circuit.

If the existing outlet lies some horizontal distance away from the new outlet, the job is more complicated. Sometimes you can drill up from the basement into the space between the studs. Drop your cable from the new outlet into the basement, run the cable along the basement ceiling, then push it up between the studs flanking the existing outlet.

In other cases you can drill down from an attic into the space between the studs, running your cable along the attic floor. This method is usually most practical when connecting the new outlet to the branch circuit by way of an existing light fixture on the ceiling below the attic.

Another useful method is to remove a length of baseboard, exposing the studs. Drop the cable from the outlet box, run it horizontally across the studs, then run it up or down again to the point of connection. Finally, replace the

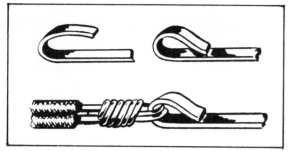

Open and closed hooks on fish wire or fish tape, also called electrician's snake.

Typical Fishing Expedition — Connecting First-Floor Switch with Ceiling Light

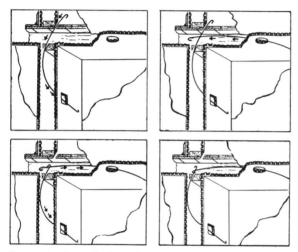

Connection is made through room above. Remove light and light box. Then cut wall opening. Take off baseboard in upper room and drill hole where wall meets floor. Drop fish wire, hooked at both ends, through that hole to switch hole. Insert second fish wire through light hole and hook first fish wire. Draw wires taut (get someone to help you). Attach cable to hook at lower opening and pull cable through from above.

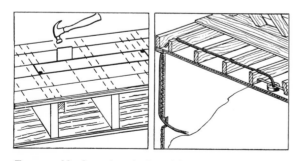

To run cable through attic floor, lift floorboards. Pry up tongue-and-groove flooring with chisel or wedge driven between boards.

baseboard. You may have to notch the studs to accommodate the cable.

Frequently, you will need *fish wire*, also known as *fish tape* or *electrician's snake* to install cable. Fish wire, which can be bought in hardware or electrical supply stores, comes in various lengths and styles. Sometimes it comes wound on a spool. The end of a fish wire has a hook that can be opened or closed by pliers. Some types, however, must be heated before the end can be bent. Make sure you have plenty of strong string on hand.

Fish wire can solve many cabling problems, depending on the layout of your home and your own ingenuity (see illustrations).

If you cannot lay your cable by any of the

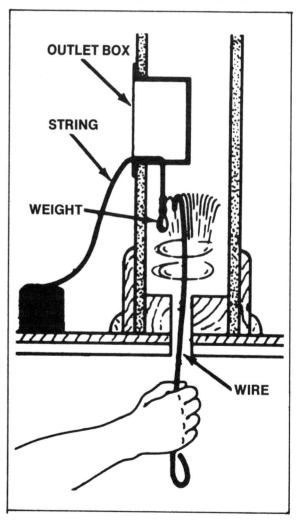

If necessary to fish from below, hang a weighted string through box. Drill a hole upward, hook weight, and pull down.

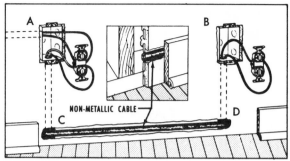

Off existing outlet (A), a new one (B) is to be tapped. Baseboard is cut away between, holes are drilled at (C) and (D), wire is fished up to both boxes. Plaster is chiseled out for recessing cable. Reinstalled baseboard (inset) conceals cable and wounds.

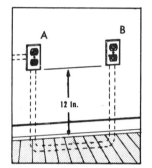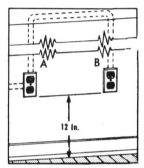

A straight line isn't always the easiest way between points. Left: *new outlet (B) is wired through basement to existing outlet (A)*. Right: *sometimes the easy way is to go up and over through the attic*.

Outlet-equipped strip for surface wiring.

methods described above, you have only two alternatives: (1) break wholesale into your walls, run the cable where you want it, then patch the walls as best you can, or (2) use surface wiring.

Plug-in Strips and Raceway: Approved surface wiring consists of insulated strips that can be plugged into any outlet and then run along exposed surfaces to any point you wish. These strips come with screws to fasten them to walls, baseboards, etc. Most types have prewired built-in outlets placed at intervals, but one type can be fitted with special outlets that can be placed along the strip wherever you wish.

These strips have plastic housing. They are not supposed to be installed closer than two inches to the floor. They should not be used

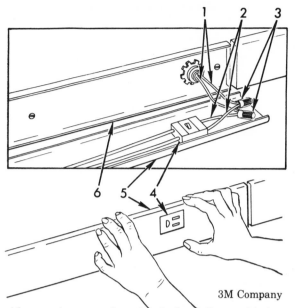

3M Company

After metal raceway channel is fixed to wall, use solderless connectors (3) to hook cable wires (1) to raceway wires (2). Snap outlets (4) into position. Finally, snap outer channel of raceway (5) over inner channel (6).

where wet conditions, such as in a cellar or laundry room, may occur. They must not be mounted near or on metallic surfaces. Most localities prohibit running such strips through a wall, floor, or ceiling.

Metal raceway is a more elaborate kind of strip. Although it is classified as surface wiring, you can substitute it for your baseboard and run cable behind it. Thus, you can have a hidden wiring job. Instead of simply plugging the connection from the cable into an existing outlet, you must wire it permanently into the outlet box. Similarly, although lengths of metal raceway come already fitted with outlet receptacles, you must make your own wiring connections.

Metal raceway can act as a continuous ground running back to your service entrance. If you decide to install metal raceway, it is best to use a type fitted with receptacles for three-prong plugs. When wiring, fasten the green wire of the receptacle to the raceway metal. Make sure that the metal connects with the grounding arrangement in your home. One way is to ground the metal to existing outlet box from which new outlets will draw current.

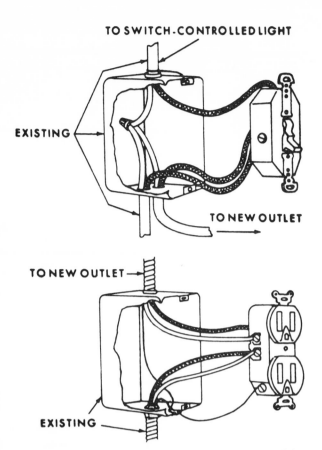

TO SWITCH-CONTROLLED LIGHT

EXISTING

TO NEW OUTLET

TO NEW OUTLET

EXISTING

Cut-away view of connections to new outlet from switch box (top) and from outlet box.

Connecting the Wires

Mechanical Hook-up: When you connect the wires for your new outlet, you are linking that outlet to the circuit from which it will draw current. The initial procedure is a mechanical one. Firmly attach the cable to the box. (Unless the connection is firm, the cable may work loose later, causing trouble behind your walls.) All boxes contain circular areas called *knockouts,* that can be punched out with a screwdriver. The round holes that result take standard fittings called *connectors.* They permanently lock the cable to the box.

You are probably using either armored BX cable or a nonmetallic sheathed cable, such as Romex. Each type requires its own style of connector. Before making the connection, it is necessary to free six to eight inches of the wires within the cable and bare about ¾ of an inch insulation from their ends. (See Chapter 4.)

Electrical Hook-up: Once the mechanical connections are completed, you can proceed with the electrical ones. If you are linking the new outlet to an existing one, the procedure is simple. Fasten the cable's black wire by looping it under the brass-colored terminal of the existing outlet. Fasten the cable's white wire by looping it under the silver-colored terminal of the existing outlet. Turn the screws tight. If you are using 3-wire grounding cable, attach the third wire to the box by a screw or clip. Finally, take the same steps at the other end of the cable to complete hooking up the new outlet.

If you are linking the new outlet to an exposed junction box, begin by unscrewing the

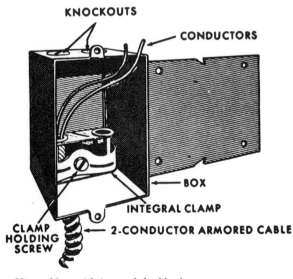

KNOCKOUTS

CONDUCTORS

BOX

INTEGRAL CLAMP

CLAMP HOLDING SCREW

2-CONDUCTOR ARMORED CABLE

View of box with integral double clamp.

Three-wire nonmetallic sheathed cable fixed to box by nonintegral fitting. Note clip holding the ground wire.

cover. You will see white wires fastened to white and black wires fastened to black by means of solderless connectors. Unscrew the connector caps. Twist the bared black wire from the cable around the black wires in the box. Twist the bared white wire from the cable around the white wires in the box. Replace the connectors; if they prove too small get bigger ones. If you are using 3-wire grounding cable, attach the third wire to the box by a screw or clip.

If you are linking the new outlet to a wall or ceiling light, you must remove the light fixture to get at the box behind it. Remove any hanger or bracket that might be in your way. You may already have done this if you threaded the cable through to the box by the fishing method.

Now, punch away a knockout. Firmly connect the cable at the hole with a proper fitting. Then, twist the bared black wire from your cable around black wires in the box. Twist the white wire around white. Replace the solderless connectors. Do not splice your black wire into the black or red wire that is the return lead from the light switch. If you hook your new outlet up to this wire, the light switch will turn the outlets on and off.

A word of caution. Bare only enough wire to fit around a screw terminal or under a solderless connector. If any bare wire peeks out from under a connector, do the job over again. Some careful amateurs anchor the cap to the adjacent insulation of the wires by wrapping a strip of electrical tape around the whole bundle.

Adding a Branch Circuit

It may be that none of the branch circuits in your home have enough unused capacity to justify tapping into them. Or, the branch circuit may be impossible to reach by boring through walls and fishing. If so, your only alternative is to install one or more new branch circuits.

Adding a branch circuit is the best method of solving an overload problem. But taking off an additional branch circuit from a fuse or circuit-breaker service entrance is not a job for the average householder. Better leave it to a professional. Some entrances have take-off connections; some don't. Some installations re-

quire additional switches and additional boxes containing fuses or circuit breakers wired to the main box. Sometimes the 2-wire system feeding the house does not have the capacity to support an additional branch circuit; a 3-wire system has to be installed. These considerations tend to involve too many technicalities for the amateur.

But having an electrician put in an additional circuit and distributing it through your home can run to great expense. What you can do is compromise. Have an electrician make the takeoff from your service entrance. Let him lead the new circuit to a conveniently located junction box (preferably in the basement). Then take over the rest of the job yourself.

Of course, your work, as well as the electrician's, will have to satisfy the requirements of your local electrical code. You may have to obtain a permit in advance and then subject the work to inspection. But running the circuit to where you want it and hooking it up to outlets or lights is not difficult. Simply follow the step-by-step procedures given under Hooking into a Branch Circuit (see page 31). From the circuit's junction box, lay your cable to where you can push or fish it to the source of overload — for example, your kitchen. You can then install additional outlets.

Installing Improvements

No doubt it has occurred to you that the techniques for getting rid of an overload can be used for putting in improvements such as additional outlets, lights, and switches. If your present branch circuits have the wattage capacity, these improvements can be tapped into an existing run of cable.

This is most conveniently done at an exposed junction box. If an exposed junction box is not available, tap into an outlet box or a light box. Use solderless connectors where necessary.

Adding Outlets to an Existing Run: The difficulty with adding outlets to an existing run is not an electrical one. The problem is that getting the cable from the old outlet to the new one may be a task. Drilling, fishing, and fastening the new box in place were detailed earlier in this chapter. To repeat the hook-up:

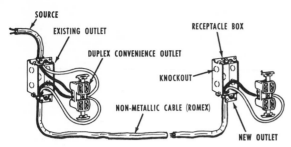

Tapping existing outlet for current to new outlet, using 2-wire Romex.

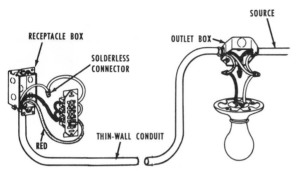

Tapping existing light for current to outlets but retaining light switch. Conduit (see Chapter 4) is shown but cable will do.

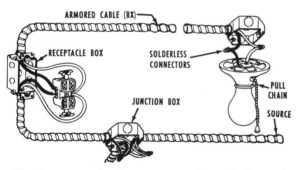

Tapping junction box for current to outlet and light. Excellent for installation from basement.

Connect the black wire of the cable to the brass terminal of the existing outlet. Connect the white wire of the cable to the silver-colored terminal of the existing outlet. At the other end of the cable, connect the black wire to the brass terminal of the new outlet. Connect the white wire there to the silver-colored terminal of the new outlet. Connect ground wires, if any, to the boxes. This completes the electrical connections.

Double Outlet Plus Switch: When you wish to tap a light box for current to an outlet, you can put in or retain a wall switch for the light at the same time. Three-wire cable can be used for the connection. Some localities, however, require that conduit be used for all improvement work. Although conduit is more difficult to get into place, it is the ultimate in safety (see Chapter 4). It is put into position first. Then through it are pulled, pushed, or fished a black wire, a red wire, and a white wire.

Connect the black wire to the brass-colored terminals of the outlets and either terminal of the switch and to the black wire in the light box. Connect the red wire to the other terminal of the switch and the black wire of the lighting fixture. Connect the white wire to the silver-colored terminals of the outlet and the white wire of the fixture. Now the switch will control the light, but the outlet receptacles will remain on duty whether the switch is on or off.

Outlet Plus Light: This combination is especially recommended when the circuit can be tapped from an exposed junction box, often found in a basement. It is easiest to use 2-wire armored or nonmetallic sheathed cable that runs up from the basement between the studs to the new outlet and new light. Installing a box for the light is similar to installing an outlet box.

Run the cable over a ceiling by lifting the floorboards. Framing under the floor should be notched out and then an opening should be cut in the ceiling to accommodate the light box. In this installation a light switch can be substituted for or added to the outlet.

Switch to Operate Existing Light: By using 2-wire armored cable (BX) or nonmetallic cable, you can readily hook up this circuit. But remember that switches are never inserted in the neutral or ground side of a circuit; the white wire is misleading. So paint the exposed white insulation black or red after you complete your connections.

Otherwise, someone working on the box at a later date might not realize that the wire was hot — and might suffer a shock.

Three-Way Switches: These handy devices enable a wall or ceiling light to be turned on or off from two different points. For example, a

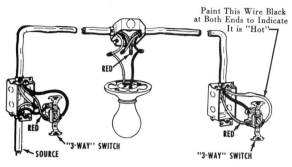

Installing 3-way switches to operate light from separate points. Three-wire cable should be used.

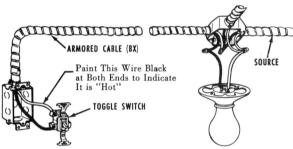

Adding switch to existing light.

hallway light can be switched on as you enter at one end and switched off as you leave at the other end. When you start up a stairway, you can switch on a light; when you reach the top, you can turn it off. Thus steps are saved, and accidents in unlit areas are avoided.

A light can be controlled by three or four switches placed at different points. More usual, and more feasible for installation by the amateur, is a light controlled from two points — for which two 3-way switches are needed.

Make your connections with 3-wire cable. Run the white wire from the current source directly to the white wire of the lighting fixture. Run the black wire from the source to a dark "common" terminal of the first switch. At

the second switch, connect a black wire to the common terminal; connect the other end to a black wire of the lighting fixture.

You should now have two unused brass-colored terminals on each switch. Your next step is to connect the two red wires in the light box. The other ends of the red wire connect to a terminal on each switch. Connect the black wire of the cable to the remaining terminal of the first switch, and to the *white* wire in the light box. The other end of this white wire connects to the remaining terminal of the second switch. Paint this run of white wire black at both ends to indicate that it is hot. The other run of white wire, connecting the first switch to the light, also should be painted black.

Cable or Conduit?

Thus far a number of types of cable and conduit have been mentioned. All have their uses, often predetermined by code regulations. For the do-it-yourselfer who plans to tap into an existing circuit and cut his way through walls, the flexible types are the easiest to work with. Some applications of armored cable and non-metallic sheathed cable have already been shown.

In general you can use either kind of cable. Thin-wall conduit can be useful because its flexibility permits it to be pushed readily into wall openings, up and down between joists, through drilled holes, and so forth. But after it is in place, wires must still be pushed or fished through it. Sometimes it is feasible to drop string through a length of thin-wall, notching it at one end so that the string can be anchored. Then, when the conduit is pushed into place, possibly you can get hold of the string and use it to pull the wire through.

For more about wire, cable and conduit, see Chapter 4.

4
Choosing and Using Wire

The primary material of all electrical circuits is wire, usually copper wire. In home electrical systems, all wire not enclosed in a metal box must be protected by something else — a metal winding, a nonmetallic sheath, or a metal conduit. Electrical codes, whether national or local, require such protection. Without it, tension, vibration, or inadvertent contact with an edged tool might rip the wires; insulation might fray or become unduly damp. Thus shorts could creep into a circuit, creating shock and fire hazards.

Kinds of Cable

When wires are gathered together within heavy-duty insulation and a protective casing, the bundle is known as *cable*. Most home repair or improvement work, whether done by amateurs or professionals, generally uses cable. Its great advantage is flexibility. It can be pushed, pulled, or fished around bends and corners; poked through relatively small holes; and make long and tortuous runs without joints or fittings (except at the electrical boxes where connections are made).

Three types of cable are in general use: (1) flexible armored cable (commonly called BX — a trade name), (2) nonmetallic cable or Romex (also a trade name) and (3) heavy-duty plastic-covered cable. Flexible armored cable is probably the most popular type for indoor wiring.

Working with BX

BX cable is composed of two or three color-coded insulated wires. The wires are wrapped in spiral layers of tough paper. The layers are then enclosed in a protective spiral winding of galvanized steel. It particularly lends itself to use in areas of the home where dampness is not a problem.

An advantage of BX is that even in its 2-wire version — comprised of a hot wire and a neutral wire — the steel winding can serve as a connection to ground for junction boxes, outlets, lights, and everything else in a circuit. BX is particularly recommended for homes with trustworthy basic grounding, for example, where connection to a municipal water system is available.

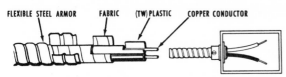

FLEXIBLE STEEL ARMOR   FABRIC   (TW) PLASTIC   COPPER CONDUCTOR

BX armored cable is omnipresent in most electrical installations. It is fairly flexible, clamps tightly to electric boxes.

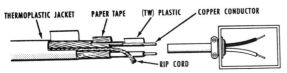

THERMOPLASTIC JACKET   PAPER TAPE   (TW) PLASTIC   COPPER CONDUCTOR

RIP CORD

Modern plastic-covered cable also is flexible — and better than BX for damp or outdoor installations.

Working with BX Cable

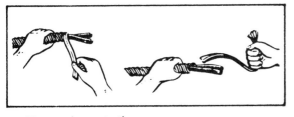

1. *Cutting off the armor.*

2. *Unwrapping protective paper.*

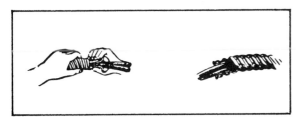

3. *Inserting anti-short bushing.*

4. *Attaching connector to cable and then to box.*

BX cable is unsuitable for outdoor or underground use. Nor should it be used where damp conditions prevail. Otherwise you can rely on it, even when it is permanently encased in masonry or plaster.

Baring the Wires: A hacksaw or heavy snip will cut right through armored cable. So you will have no trouble cutting it into the lengths you need.

A bit more tricky is baring the wires in the cable in order to make connections. It is best to cut off enough armor to expose from 6 to 8 inches of the insulated wires. Use a fine-tooth hacksaw to remove the protective spiral-wound steel armor. Hold the saw at a right angle to the spiral rather than at a right angle to the cable length. Carefully cut through the armor, making certain that the saw does not strike the insulation of the wires beneath. The cut does not have to be an extensive one; saw just enough to pierce the armor.

Next, place a hand on each side of the cut and grasp the cable. Bend it back and forth to enlarge the cut; then twist the shorter end sharply. The armor should snap right off. If it doesn't, bend the cable back and forth a few more times until the armor parts.

Now, using a knife, slice off enough insulation to bare the wire ends.

Bushings Are Important: Electrical codes mandate that a protective fiber bushing be inserted at the cut end of armored cable. Otherwise the raw edge of the cut armor might penetrate the wire insulation and cause a short circuit. Bushings are available wherever cable is sold.

To create room for the bushing, unwrap a few turns of the protective paper covering the wires. Pull the wrapping sharply so it will tear off inside the armor; then insert the bushing.

Connecting BX to Boxes: As mentioned earlier, where BX joins a junction box or other electrical box, the cable must be firmly anchored. You can buy boxes with built-in clamps for gripping armored cable. Then all you have to do is insert the cable into the clamp and tighten the clamp screw.

If the box you buy does not have clamps, use a cable connector. After the bushings have been inserted, slip the cable through the connector. The threaded end of the connector

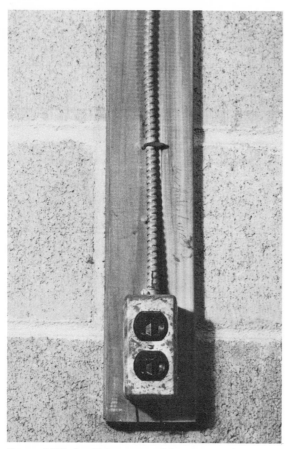

Exposed BX should be supported by straps or clips, as shown in this basement installation.

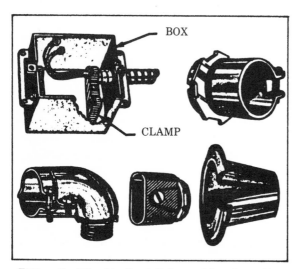

Fittings for BX cable. Top left: *integral box clamp*. Top right: *nonintegral box clamp for single cable*. Bottom left to right: *sharp-corner guide for single cable, box clamp for two cables, anti-short fiber bushing.*

should face the box. Tighten the holding screw, securing the connector to the cable. Punch a knockout slug out of the box. Insert the threaded end of the connector through the hole. Turn the connector locknut tight on the threads, assuring a lasting and well-grounded connection. To tighten the nut, seat a screwdriver against a cog and tap with your hand, forcing the nut counterclockwise.

In new construction and exposed work, BX should be fixed in place by straps or staples every 4 feet of the run and 6 to 12 inches from any electrical box. Most local codes do not require this when the work is being done in an existing home and the runs are concealed.

BX Fittings: You can purchase a variety of accessories from your electrical supply dealer, making the work of installing BX easier and faster and helping to make the finished job neat and safe. These fittings can be important in solving special problems. For example, a duplex connector permits entry of two cables through one box knockout. Angle connectors, rounded side connectors, and end fittings are also available, the latter permitting you to change from cable to noncable wiring.

Nonmetallic Sheathed Cable

Sturdy and flexible, the Romex type of nonmetallic cable is well-suited for repair and installation work. It usually costs less than other common cable of equal electrical specifications. And despite its strength, it is light in weight. But Romex has one disadvantage. Being nonmetallic, its sheath does not ground from box to box. To assure grounding, 3-wire nonmetallic cable should be used. The third wire should be used to ground the cable to each box it encounters.

To bare the interior wires, slit the braided covering of Romex with a cable ripper or an electrician's knife. Cut parallel to the wires for a distance of from 6 to 8 inches, making certain that you do not damage the insulation covering the wires. Snip off the slit outer braid; then cut off enough insulation to bare ¾ of an inch or so at the ends of the wires.

Box Connections: Like BX, nonmetallic cable requires mechanical junction with boxes. Simi-

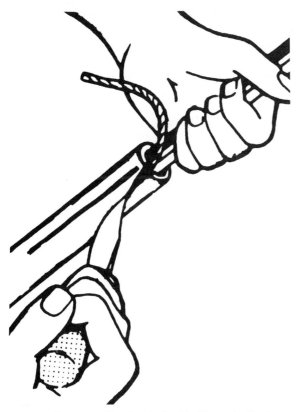

To install nonmetallic sheathed cable (Romex), slit the exterior braid for a distance of 6 to 8 inches. Take care not to slice into insulation of interior wires.

lar connectors are used except that in cross section they are oval rather than round. You can buy boxes already fitted with connectors, or you can buy the connectors separately. Codes require prefitted boxes to contain a threaded grounding hole so that you can lock the ground wire in place by turning a screw. If you buy the connectors separately, be sure that you also buy suitable clips for locking the ground wires to your boxes.

The rest of the procedure is the same as for BX cable. Slip the connector over the cable end and fasten it into place with two holding screws. Push the connector into the knockout hole of the box and secure it by turning the locknut tight. Then use a clip or screw to anchor the ground wire to the box.

Do's and Don'ts: Because of its "soft" sheath, Romex may tempt its user to make a direct splice in the middle of a run, cover it with tape and let it go at that. By code, this is not permissible. As for any type of home wiring, all splices and connections must be made within boxes only. So run the cable from box to box in unpatched lengths.

Do not bend nonmetallic cable too sharply, or in the course of time, the internal wires may break and cause an open circuit.

Where such cable may be subject to abrasion or impact, as in a home workshop, protect it by running it through metal conduit.

Wherever practical (for instance, on studs and joists), use straps to support the Romex. These are sold by electrical supply retailers. Strapping is required even if the wiring is to be concealed. However, where the cable is fished through a wall, floor, or ceiling, you may omit the straps.

Do not bury Romex in plaster or masonry. Do not bury it underground. It may be used for either exposed or concealed work but only use it indoors.

Exposed Cable: Romex is often used for wiring attics, basements, garages, and other locations where it will not be concealed within walls. In such cases, it is best to give it solid support. On vertical runs, for example, the cable should be strapped to a stud. Where the cable runs horizontally or at any other angle across studs or joists, it should be strapped along 1 × 2 or 1 × 3 running boards affixed to them. It is also permissible to run the cable through holes bored in overhead joists.

In unfinished attics, cable can cross floor joists only if it is supported by a running board or guarded by protective boards erected on both sides of the cable to form a channel. Romex must be protected similarly when it crosses rafters less than seven feet above the attic floor. If your attic does not have a permanent set of stairs, skip the running boards or guard boards except within six feet of the entrance.

Plastic-Covered Cable

Cable encased in tough plastic is a relatively new development. This type of cable is unusually versatile. Its exterior insulation is heavier than that of Romex, which it otherwise resembles. The covering resists rot and fire, as well as acids, moisture, and impact.

For those reasons, heavy-duty plastic-covered cable may be run through holes in brick, masonry, or plaster walls. It can be used both indoors and out. It is especially useful for wet or corrosive locations and for running underground between buildings or to serve outdoor lights.

Conduit for Protection

Electrical conduit, which resembles water pipe, is the surest safeguard against damage to wiring systems. Conduit is found more commonly in commercial buildings than in homes. One reason is that conduit is more difficult to work with than cable. Also, conduit is usually more costly.

After the conduit is installed, the necessity of drawing the needed wires through it remains. Some types of cable may be drawn through it also. When repairing or refurbishing an existing house, the chief problem with conduit is getting it installed. Once it is in place, pushing or fishing the wire through is relatively easy.

Thin-Wall Conduit: Thin-wall conduit is easier to cut and bend than heavier rigid conduit. It is almost universally accepted by electrical codes for general wiring, and some codes require it in new construction. Installation to boxes is not difficult, since the joints and connections are made with special threadless fittings.

Known also as EMT, thin-wall conduit can be used for both concealed and exposed work, either indoors or out, and in a wet or dry location. However, it should never be buried in cinders or cinder concrete.

The wires used in EMT work are single strand. They are covered with thermoplastic insulation and color-coded. Because the conduit is metal, it makes a continuous ground between boxes.

Special fittings, resembling plumbing couplings, connect one length of thin-wall conduit to another. A 90-degree elbow connector is available for going around sharp corners.

Rigid Conduit: The sizes and dimensions of rigid conduit are the same as those of standard stainless steel pipe. It comes with a black enamel finish (for indoor use only) and with a

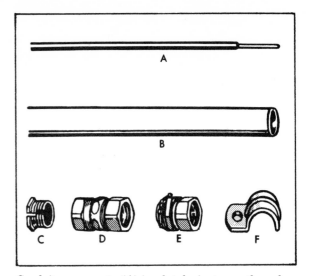

Conduit components: (A) insulated wire to run through conduit, (B) thin-wall conduit, (C, D, E) connectors and locknut for attaching conduit to box, (F) strap for securing conduit to house framework.

galvanized finish (for indoor or outdoor use). Like other conduit, it has a smooth inside surface so that wires may be passed through without too many problems. The conduit is annealed to permit bending, but this does not mean it is easy to bend! Amateurs should not try to use rigid conduit except outdoors or in readily accessible locations such as garages.

Flexible Conduit: Generally called "Greenfield," flexible conduit looks like the protective steel casing of armored cable. Greenfield is larger in diameter, however, permitting wires to be pushed or pulled through it. While not extensively used for general home wiring, Greenfield is excellent for jobs where extra protection of the wire is needed and thin-wall or rigid conduit is impractical to install.

For example, concealing either thin-wall or rigid conduit in a finished wall might mean removing a large amount of lath and plaster — and then replacing it all. But Greenfield conduit can be inserted through any small hole in the same way as flexible cable. Often the only openings needed will be those you will use later for outlet boxes. The only difference between installing flexible cable and Greenfield is that the latter requires the wires to be pulled through afterward.

Cutting Conduit: Use a hacksaw to cut both

thin-wall and rigid conduit. Rigid conduit can also be cut with an ordinary pipe-cutter.

Ream the cut ends to remove sharp edges or burrs that might damage wires as they are pulled through. Remove outside sharp edges with a file so that the fittings will slide on easily. Greenfield can also be cut with a hacksaw, but do not try to bend or twist it apart as you would armored cable.

Bending Conduit: To get around obstructions or change direction, it is necessary to bend the conduit. To figure out the exact bend required in advance is rather difficult even for professionals; your best recourse may be trial and error.

The practical way to bend thin-wall or rigid conduit is with a proper bending tool. Make bends gradual. Take care not to kink or collapse the conduit, or it might be difficult afterward to get your wires through. No more than four quarter-bends or the equivalent should be made in a single conduit run; the fewer the bends, the easier it will be to thread the wire.

You do not need to make bends if you are using Greenfield conduit, since it is completely flexible.

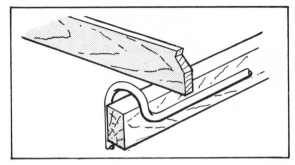

Instead of going through ceiling beam, bend conduit to saddle it.

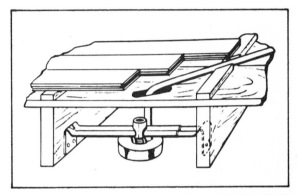

Manner of bending conduit for installation of ceiling light in first floor room. Note notch in sleeper to contain conduit. Hanger bar fixed to joists holds box in position.

previously mentioned. Fit the threadless end of the connector over the conduit and tighten the compression nut. Then insert the threaded end through the knockout of the box and tighten the locknut securely against the inside of the box wall. Thin-wall conduit can be connected to rigid conduit by means of a threaded adapter that screws into a standard pipe coupling.

Rigid conduit can take standard pipe fittings, but first it must be threaded with the same tools used for water pipe. Connectors are available to couple it to boxes.

Greenfield comes with its own special connectors. These connectors are similar to those used for connecting armored cable.

Before pulling wires through the conduit, the conduit should be fastened in place with a pipe strap every six feet on exposed runs and every ten feet on runs that will be concealed. The conduit should then be connected to its junction and other boxes. Wires must run continuously and in one piece inside the conduit; splices and

Conduit bender (top) and manner of use.

Making Connections: Thin-wall conduit is attached to boxes with the special connectors

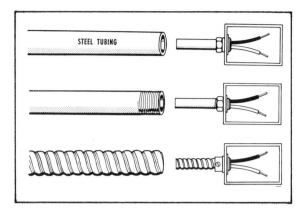

From top down: *thin-wall, rigid, and Greenfield conduit.
Each must be fixed to boxes by appropriate fittings.*

electric connections must be made only inside
boxes. In a short run, the wires can simply be
pushed through. If this cannot be done, insert a
fish tape, hook the far end to the wires, and pull
them through. In a 2-wire circuit, use a white-
colored wire for the neutral or ground and a
black-colored wire for the hot side. In a 3-wire
circuit, include a white, a black, and a red wire.
All of the wires should be pushed or pulled
through the conduit at one time.

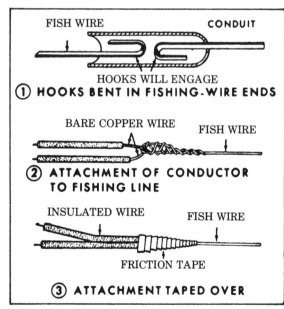

*Three ways to pull wires through conduit. Same fastenings
can be useful for pulling through standard cable when conduit is not used.*

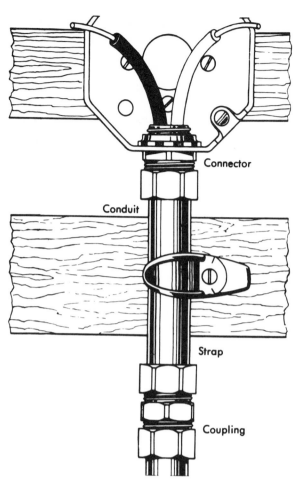

*Finished installation, showing how conduit is coupled,
strapped, and connected to box.*

The Wiring Connection

One of the most common jobs you'll run
into is fastening wires to screw terminals on
outlets, switches, and other electrical
equipment. Such connections occur either at
the end of a wire or somewhere along its
length.

Start by stripping about ¾ inch of wire
clean of insulation. If the connection is at the
wire's end, loop it under the screw in a
clockwise direction; tighten the screw to
secure the loop firmly. If the connection is at
some point along the run of the wire, do not
cut through the wire. Simply loop the bared
but continuous portion around the screw;
then tighten the screw.

In junction boxes and other electrical boxes, the best way to connect wires is by baring their ends, twisting the ends together and then inserting them in a solderless connector. Turning the connector clockwise locks the ends tight and at the same time insulates them. The same effect could be achieved by soldering the joined ends and covering them with insulating electrical tape; to do that you would have to work with a hot iron in a limited, inconvenient space.

Actually, the connection within the box is a splice of two or more wires. Since there is no mechanical strain on such a splice, it is permitted by electrical codes. But electrical codes do not permit splices of any kind, whether by connector or solder, outside of a box.

Nevertheless, occasions arise when your only recourse is to splice. For instance, you may need to make a temporary emergency repair. Or you may want to work on wiring independent of the house circuits — for example, a broken lamp cord, a low-voltage chime circuit (see Chapter 5) or the wiring of an electrical toy. The most popular and one of the best ways for joining the ends of two wires is the so-called telephone splice. Cut about 3 inches of insulation from each wire. Scrape the bare wires with sandpaper to remove oxidation. Cross the wires about 1 inch from the insulation and wind each wire around the other some 6 or 8 times.

For a good electrical connection, splices should be soldered. Use resin-core — not acid-core — solder. Apply the soldering iron to the bare wire — not the solder — until the wire gets hot enough to melt the solder. If the heat melts the insulation before it melts the solder, you need a larger soldering iron.

After soldering, insulate all bare wire by wrapping it in plastic electrical tape. Then add mechanical strength by wrapping the plastic in friction tape. The wrappings should overlap the original insulation of the wires.

WARNING: To heat up an electric soldering iron you plug it into a live circuit.

So if you work with that iron on the wires of another house circuit — even a dead circuit — you risk shock. Unless you are sure you have a type of an electric iron specially designed to eliminate this hazard, stick to an old-fashioned iron that is heated in a flame.

Frequently, soldering can be avoided by splicing with solderless connectors. If you are tapping into a continuous length of one wire with the end of another wire, you can use a Scotchlock tap connector. This even saves you the trouble of stripping the wires.

One word more: When removing insulation, taper it as you would when sharpening a pencil. That way you avoid nicking the wire and weakening it mechanically.

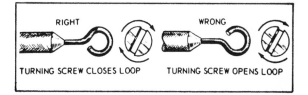

RIGHT   WRONG

TURNING SCREW CLOSES LOOP   TURNING SCREW OPENS LOOP

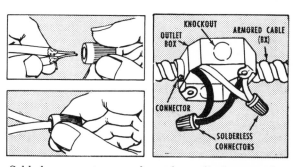

KNOCKOUT   ARMORED CABLE (BX)

OUTLET BOX

CONNECTOR

SOLDERLESS CONNECTORS

Solderless connectors can also make a splice but properly should be used only within electrical boxes. Connectors come in various sizes. They should be large enough to completely contain the bared wire ends.

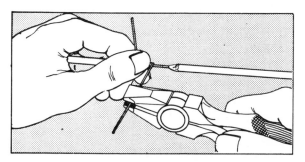

Making a telephone splice. Wires should be wound tight.

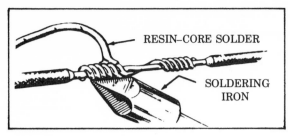

To solder for good strength and electrical conductivity, apply soldering iron to splice. Soldering iron is being applied to the wire, not the solder, in soldering this telephone splice. Method assures good strength and electrical conductivity.

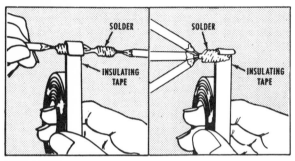

To insulate splice, wrap all bare wire in plastic electrical tape.

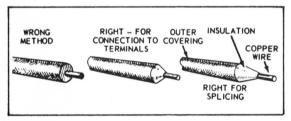

To avoid nicking and weakening wire, cut off insulation at a bevel.

Selecting the Wire Size

Electrical wire is manufactured in certain standard thicknesses or gauges. For any wiring job, it is imperative that you choose not only the right *kind* of wire but also the right *thickness*. If the wire you choose is not heavy enough to carry the amperage load intended for it, an overload will result. Most wire approved for home circuits functions at a maximum temperature of 140 degrees Fahrenheit. A few heavy-duty types can handle a maximum temperature of nearly 170 degrees.

Avoid replacing a wire with another wire of a smaller gauge. You can replace wire with other wire of a larger gauge. This can be a safety measure but also considerably more expensive.

After you have decided on the maximum wattage you want a new branch circuit to carry, you can select the proper wire size from the table below. To translate the table into the maximum watt load the wire will take, multiply the amperage by the voltage. For example, if the table states that 14-gauge wire has a capacity of 15 amperes, it can safely handle 1725 watts (15 × 115).

Copper Wire Sizes Required for Various Circuit Capacities*	
Current-carrying capacity (in amps)	*Minimum gauge required***
15	14
20	12
30	10
40	8
55	6
70	4
80	3
95	2
110	1
125	0

For 115-volt single phase circuits, with not more than 3 wires in cable or conduit, the indicated sizes will prevent more than a 3-percent voltage drop.

**Applies to one-way run from source to load of not more than 49 feet. For run of 50 feet, use wire 1 gauge thicker. For each additional 50 feet, use wire 1 or 2 additional gauges thicker.*

However, there is a catch. The length of circuit wires, measured one way from your service entrance, may be more than 50 feet or so. The internal resistance of a wire traveling such distances causes a perceptible voltage drop, and builds excessive heat. Accordingly, for each additional 25 or 50 feet a wire travels beyond the first 50 feet, you should use wire that is one gauge thicker. For example, if a wire travels 80 feet from a service entrance and you would normally use 14-gauge wire, use 12-gauge wire instead.

Unless you are using conduit, your wires will

be encased in cable. A cable containing two 14-gauge wires is called a 2–14 cable. The cable does not alter the characteristics of the wire it contains.

A smart thing to do when you buy wire, in cable form or otherwise, is to consult a reputable dealer. Tell him exactly what you want to use the wire for, and he will help you select the appropriate style and size.

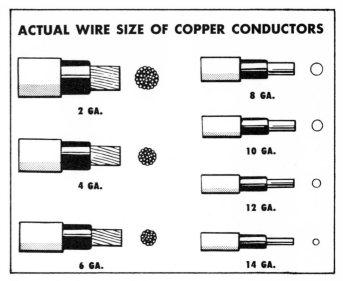

Copper wire shown actual size.

5
Lights, Doorbells and Chimes

A short or an open circuit may mean that one of the light fixtures in your home must be repaired. Or maybe you know you have to do something about the fixture because the darned thing just won't work. Replacing an outlet or switch is inexpensive, but replacing a fixture can be quite expensive. Therefore, you might prefer to repair rather than replace a fixture.

The interior of a fixture is not as easy to get at. First you must dismantle and dismount it. Yet there is nothing really complicated about dismantling a light fixture.

Ceiling Lights

Since the bulbs in many ceiling fixtures cannot be changed without taking off the fixture's glass enclosure, no doubt you already know how to remove the enclosure. This exposes one or more bulb sockets and a canopy — the plate that covers the electrical box sunk into the ceiling. Test the bulb. If the light fixture did not fail simply because of a blown bulb, you will have to investigate further.

WARNING: Before dismounting a light fixture, eliminate any chance of shock. Cut off all current by throwing your main switch or main circuit breaker to *off*. Remove the fuse protecting the circuit, or move its circuit breaker to the *off* position. If it is necessary to keep other circuits going in your home, now you can return the main switch or main circuit breaker to *on*.

Assuming you are sure that any wall switch controlling the light is working properly, the next step is to remove the screws (or nuts) binding the canopy to the box in the ceiling. This exposes the connections from the box to the light; connections, actually, that wind up at the bulb sockets. Closely examine all wires and their connections. Frayed insulation may have given rise to a short; if so, wrap the frayed spot in electrical tape or replace the defective wire. A solderless connector may have worked loose; if so, tighten it. A wire may have broken beneath its insulation (you can check this with the low-voltage continuity tester mentioned in Chapter 1); such a wire, of course, must be replaced.

If none of the above measures fixes the light, suspect the bulb socket. If there is more than one socket and one does not work although the others do, that's a giveaway. Try to pry up the center spring contact within the dead socket with a screwdriver; sometimes a contact does not rise high enough or has been pushed to one side, so that it does not meet the bottom contact of the bulb. If this does not work, you will have to replace the bad socket.

In most fixtures the socket can be removed by unscrewing it from the bracket holding it. Take it with you to the store and buy an exact duplicate. Install it, reconnect the wires, and reinstall the light fixture. Some sockets have screw terminals for connections; others come with wires permanently fastened within the socket body, leaving the other ends of the wires to be fastened directly to the box wires by means of solderless connectors.

Always join black wires to black and white wires to white. Make sure the bared wire ends are completely covered by the connector cap.

Pull Switches: Some ceiling fixtures are operated by pull-chain switches. When these fail, they can be removed from the fixture and then replaced with a switch of the same kind. If the chain has been broken off, it too is replaceable. It is held to the switch by a ball-housing that can be pried apart with a knife and closed with pliers. To substitute a wall switch for a pull-chain type, remove the connections to the pull-switch, bind them in a solderless connector, and then remove the switch.

Wall Lights

There are only a few differences between ceiling fixtures and wall fixtures. First, the canopy or plate covering the electrical box of a wall light is generally easier to remove and the internal mechanism easier to get at. Usually all you have to do is lift off the globe or shade and turn a couple of decorative nuts.

Second, a wall light may be controlled by a turn-switch or a push-switch. You can loosen these types of switches by removing a single

nut. Hence, they are not difficult to replace. If the switch is built right into the socket, it is easiest to replace the whole socket with another containing a similar switch.

Replacing Fixtures

If your old lighting fixture is a hopeless mess — or you are tired of it — you may want to replace it with a new one. Fixtures commonly come with prewired sockets. All you have to do electrically is join the socket wires, and switch wires if any, to the wires in the box. Use solderless connectors. In most fixtures one wire is white; the other black. Sometimes wire intended to go to a switch is colored red.

There are also fixtures that contain wires of the same color, except that a colored tracer thread is interwoven into the outer insulation of one. Always connect the tracer wire with the white wire in the box. The partner of the tracer wire, after being connected to a black (or red) wire in the box, should then itself be painted black.

As you can see installing a fixture is not difficult electrically. But you may have a problem in mounting the canopy to the box securely enough to suspend the fixture. There are a number of ways for suspending a light fixture. Which one you use depends on the style and weight of the new fixture and the size and shape of the box. Good fixtures often are packaged with detailed instructions for hanging them, but here are some tips on how it's done.

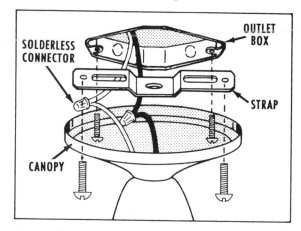

Some ceiling fixtures mount by means of a strap; screws attach fixture to strap.

Mounting to an Eared Box: If your box has ears, or threaded tabs, screw a mounting strap to them; then make your necessary wiring connections. Finally, use long screws to attach the canopy to the strap. The canopy is threaded to take such screws. The required hardware may come with the fixture when you buy it, or you can buy the screws at any hardware store.

Mounting to a Non-eared Box: If your box lacks ears, you are mounting one of the heavier types of fixtures, and you will need a stud. Your box may already have a stud. If not, you can replace the box with one that does, or buy a stud and bolt it into place (most light boxes already have holes drilled for the purpose). Invert the strap, slip it over the stud, and then fix it firmly in position with a locknut. Connect up the wiring; then fasten the canopy to the strap with screws. Again, some or all of the hardware may come with the fixture.

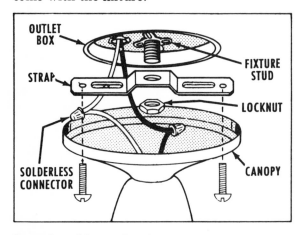

Integral or add-on studs in box mount other fixture types.

Hickey Mounting: This is useful for suspending large drop fixtures, fixtures that hang from chains, and the like. The hickey, a threaded hanger support, is turned tightly onto the stud like a nut. It is used in conjunction with a hollow nipple that passes through the center of the canopy. Wires pass from the fixture through the nipple and emerge through the hickey. A locknut secures the hickey to the nipple.

If you have an eared box but neither a hickey nor a stud, sometimes you can mount the fixture as follows: Screw a strap to the ears. Screw a locknut an inch or so down onto the

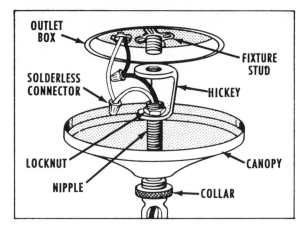

Drop fixtures usually require a hickey support threaded to screw over stud.

nipple. Insert the nipple into the center hole of the strap. It may have threads that turn into the strap, or it may be small enough to pass through the strap hole. Screw a locknut down on the nipple from the box side of the strap, and tighten both locknuts to secure nipple.

Wall Installations

The canopy or covering plate of many modern wall fixtures is not large enough to cover a round or octagonal light box. If you have such a box in the wall, make sure that you buy a fixture large enough to cover it. If your box is the narrower kind used for outlets and switches, you will have no problem.

Mount a small fixture strap to the box with

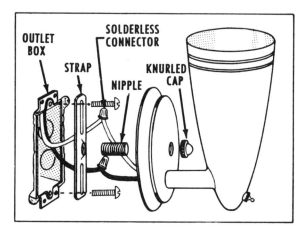

Wall fixtures are commonly fastened in place by strap and nipple. Strap screws to box.

two screws, just as you would a faceplate. Screw in the proper short nipple; then make your wire connections. Place the cover plate so that its center hole fits down over the nipple. Fasten the fixture in place with the decorative nut that came with it.

You may find that the box already contains a nipple bolted to its bottom. If so, an extension nipple or adapter must be screwed down partway onto the box nipple before the fixture nipple is inserted into the adapter's other end. Then, complete the wiring connections and fasten the fixture in place with the decorative nut.

Installing Additional Lights

The procedures described above can also be used for putting in fixtures where none existed before. However, certain preparatory steps must first be taken. These consist of: (1) *tapping a branch circuit for current to go to the new fixture* (this can best be done at an exposed junction box or an outlet box), (2) *cutting the hole for a light box where the new fixture will hang* (plan the hole for the type of box that will make it easiest to hang the fixture), (3) *fishing or otherwise getting cable to the fixture site from the tapped source* (you may prefer the source to be a wall outlet, since a switch for the light can be readily installed at that point), and (4) *installing a box at the site to contain the connections and from which to hang the fixture* (after which, you use standard connectors to lock your cable to this box and to the box at the source). These essential, and often extensive, preparations are covered under Installing Improvements, Chapter 3.

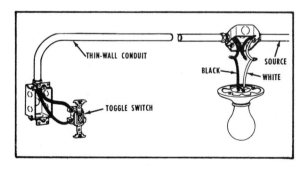

Connections for switch and ceiling light. Wires must be fished through conduit if latter is used.

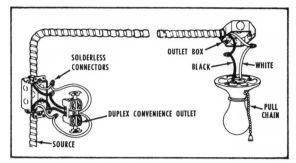

Connections for outlet and ceiling light, omitting wall switch.

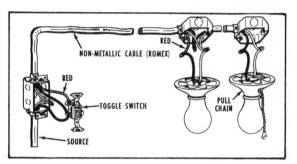

Connections for two ceiling lights, one controlled by wall switch.

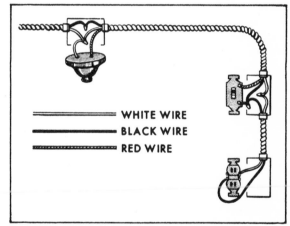

This is a basic circuit on many runs: connections for ceiling light, its wall switch, and an outlet beyond the switch. Switch controls light only, not the outlet. Never install the switch in the white-wire run.

Fixing Fluorescents

Fluorescent lights are a modern technological wonder. With continuous use, they will last from 10,000 to 20,000 hours, five or ten times longer than run-of-the-mill incandescent

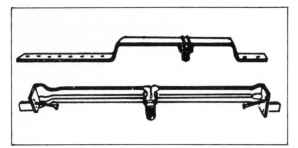

Bar hangers, purchasable in hardware stores, can affix new box to overhead joists.

bulbs. Even when turned on and off a lot, as lights normally are in homes, the fluorescent will live perhaps five times longer than the average incandescent. Further, the average fluorescent light gives off three times more light than an incandescent drawing the same wattage. That's a big consideration, considering the cost of electricity these days.

Of course, fluorescent fixtures are not impervious to breakdown. But although the way the fluorescent tube works is rather complicated, repairs are simple.

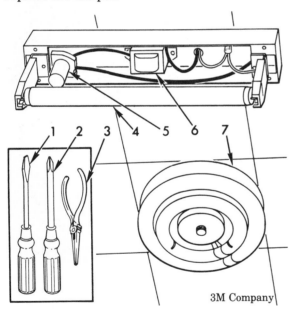

3M Company

The only tools needed for dismantling and rewiring fluorescent fixtures are screwdriver (1), Phillips screwdriver (2), and pliers (3). Most fluorescents are straight tubes (4) mounted by means of two prongs at each end. Starter (5) may be at various locations on fixture. Rapid-start fluorescents have no separate starter; latter is built into ballast (6). Some fluorescents, especially those used for kitchen lighting, are circular (7). To remove circular tube, pull straight down and it will release from sockets.

An incandescent bulb contains a filament — a strip of tungsten or other metal — that heats up when a current passes through it, giving off light. On the other hand, although a fluorescent contains small filaments at both ends of the tube, these serve only to heat up the mercury vapor within the tube. They then shut off. The electricity passes directly through the heated gas, which transfers its energy to phosphor that coats the inside of the tube. The phosphor then glows brightly.

A problem with a fluorescent light usually involves the tube itself, the starter (a small barrel-shaped device whose job it is to heat the filaments), or the sockets (at each end of the fixture) that the tube fits into. A device called a *ballast,* which works like a transformer, is also housed in the fixture. The ballast can also give trouble, but it is not too hard to replace. How to solve some of the problems that may occur with fluorescent lights is discussed below.

Tube Keeps Blinking After Lighting: Normally, when you turn on a fluorescent, it will blink a few times. But if it keeps on blinking, obviously something is wrong. It may be that the tube is not firmly seated in the sockets; so try tightening it. The best way is to start from scratch. Remove the tube, align its end pins vertically with the socket slots, insert the tube, and turn it until it clicks in.

If the light still blinks, examine the pins to see if any are misaligned and so are not making proper contact. Use a small-nosed pliers to straighten a misaligned pin. Another possibility is that the pins are encrusted with dirt. To correct this, lightly rub the pins with steel wool. Be careful — if you rub too hard you can ruin the pins.

If the above measures do not cure the blinking, a socket may be loose. **Shut off the house current to prevent shock.** Then inspect the metal housing of the fixture for screws that hold sockets in place. Tighten these screws carefully.

If the blinking still occurs, the starter is probably responsible. It is usually sunk in a hole at one end of the tube or under it, but it also may be found on the outside of the fixture or behind it. **Shut off your house current.** Then take out the starter. Most starters have the word *remove* etched on them, along with an arrow pointing left or right. Press in the starter as far as it will go; then turn it in the direction of the arrow until it pops out.

A volume could be written on how to get the correct replacement starter. It is best to go to your hardware or electrical supply dealer and show him your old starter. Maybe he carries the volume in his head and can come up with the right replacement for you.

The fluorescent may also blink because the fixture is located in a cold basement or other spot where the average temperature is less than 50 degrees. If you have an under-50 situation, shop for special low-temperature tubes.

Finally, the blinking could be caused by low voltage in your house circuit. This is something the utility company or an electrician can check for you.

It is important to take care of a blinking tube right away. Each time the tube blinks it is, in effect, restarting. Each start shortens the tube's life.

Tube Glows Brighter at Ends: When a tube behaves in that manner, the starter is the likely culprit. If a new starter does not solve the problem, the tube is probably on its way out. You might as well replace it.

Tube Darkened at Ends: Darkening starts when a fluorescent begins to go out. It still may have many hours of life left in it, so you need not replace it right away. However, if the tube was recently installed and has already darkened at the ends, suspect the starter.

Tube Discolored at One End: If the discoloration is at one end only, there is usually no cause for worry. Just take out the tube and reverse it end for end.

Tube Hums: This problem is caused by the ballast being loose. Tightening the screws that hold it on could solve the problem, but unless you are experienced with electricity, it is best not to fuss with the ballast. Sometimes the problem is less melodious — a buzz rather than a hum. The buzz can also be the consequence of a loose or defective ballast.

Rapid-Start Fluorescents

The problems and solutions listed above apply to fluorescent fixtures with starters. Another type of fluorescent is called *rapid-*

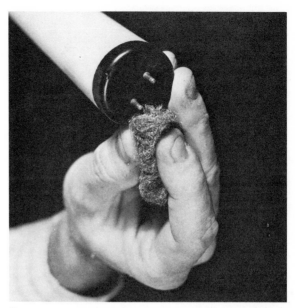

Prong terminals of fluorescent must be kept clean for good contact. Rub gently with steel wool.

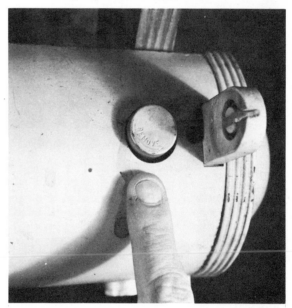

Starter unit can be snapped out for replacement. Defective starter can cause a number of troubles. Replacement starter must be of exactly the same specifications as the old one.

start, because it lights instantly. If a problem arises with a rapid-start fluorescent, you can try all the countermeasures suggested above except for replacing the starter (since there isn't any).

Avoid turning fluorescents on and off excessively. It is better to let it stay lit for an hour or

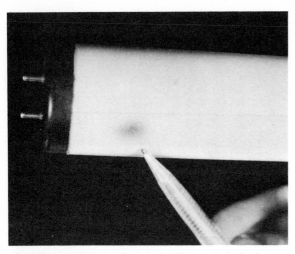

Darkening at either or both ends? If the tube is fairly new, suspect the starter. If the tube is old, it is weakening but probably will last a good while.

two than to turn it off, leave, and come back after that a short while to turn it on and off again. Finally, remember to clean the tube and its reflector with a rag once in a while. You might be surprised by the increased candle-power a clean fixture will give.

Doorbell and Chime Defects

Once upon a time, many homes used dry-cell batteries to operate doorbells and buzzers. These days transformers are almost always used for that purpose, as well as for powering door chimes. Doorbells and chimes also operate on exactly the same type of circuitry. Any circuit that operates a doorbell will also operate a chime, and vice versa. And both are activated by pushbuttons.

Four things may have gone wrong when a doorbell or chime will not operate: (1) the transformer, (2) the pushbutton, (3) the bell, buzzer, or chime mechanism itself, or (4) the wiring circuit.

Transformers: These are sturdy devices built to last a lifetime, and usually they do. The high-voltage side of the transformer is hooked up directly to your service entrance; therefore, the most common practice is to place the transformer within the entrance, alongside it, or somewhere quite near it. The transformer steps down the high voltage to between 6 and 20 volts. That lower voltage cannot shock or hurt anybody. It leaves the transformer from one low-voltage screw (or nut) terminal, operates the bell or chime mechanism, and returns to a second low-voltage screw terminal.

Check the terminals. If they are dirty or corroded, disconnect the wires, sandpaper everything clean, and then reconnect them.

If you must test the transformer, try to hook the low-voltage terminals to some low-voltage device (for example, a 12-volt automobile bulb). The transformer may blow the bulb, but if it does or if the bulb lights, the transformer is working. Another way to test the low-voltage side of the transformer is to disconnect the wires from the two low-voltage terminals. Then touch a probe of a low-voltage continuity tester to each terminal. If the tester bulb lights, the transformer is healthy.

The high-voltage terminals are usually a black and a white wire coming from within the box and feeding directly into a junction box or your service entrance. In general, leave these terminals strictly alone. They can shock you with 115 volts.

Snapit (trade name) transformer reduces 115 volts to 16 volts; other reductions are available. This transformer is a "speed-lok"; that is, it can be inserted directly through a junction-box knockout and locked there simply by tightening one screw. Visible wire terminals connect to high-voltage terminals in box or service entrance. Low-voltage terminals of transformer are on its far side in this drawing and so cannot be seen.

If you decide to install a replacement transformer, first throw your main switch or main circuit breaker to *off*. Connect the terminals of the new transformer exactly as those of the old one were connected. Only then can you return your main switch or circuit breaker to the *on* position.

Pushbuttons: You can tinker with pushbuttons by removing the cover plate and then sandpapering the contacts and fooling with the spring. But the better course is to install a replacement. This is especially true if the problem is that sometimes the bell or chimes keep ringing continuously. Make sure the wiring connections are tight.

Bell, Buzzer, and Chime Mechanisms: Bell and buzzer mechanisms depend on a make-and-break contact located on either side of the clapper arm. The exterior contact is visible after you remove the bell housing. Sometimes you can adjust it by pushing it one way or the other, getting the bell to work.

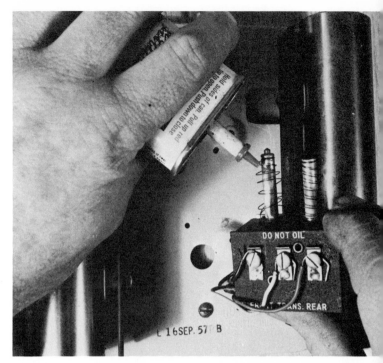

Chime strikers may stick occasionally. They are not meant to be oiled. Lighter fluid keeps them clean for smooth operation.

Two-chime unit with decorative housing removed. Wiring connections are simple; terminals should be checked for cleanliness and tightness if device fails to operate.

But the chief enemy of a bell is dirt that works its way between the contacts. Use fine sandpaper to clean away dirt and corrosion, being careful not to disturb the position of the contacts. Similarly, brighten the wires where they join the bell terminals and clean the dirt from under the terminal screws.

Faults with chime mechanisms are more difficult to analyze, since they involve springs and other secondary mechanisms that vary considerably from one to the other. If the problem is not caused by dirt or corrosion at the terminals or contacts, and if it does not yield to lubrication (provided the manufacturer's instructions call for it), you probably will have to send for a serviceman.

Wiring Circuits: Checking an entire circuit can be a nuisance and quite difficult. Often it is easiest to disconnect the old wires leading from one device to another and put in new wires. If the circuit still does not work, disconnect the old wires leading from the second device to a third one and rewire. If necessary, repeat the procedure until you have rewired the whole circuit. Take the shortest distance between points until you have located the trouble; then use longer lengths of wire for permanent installation.

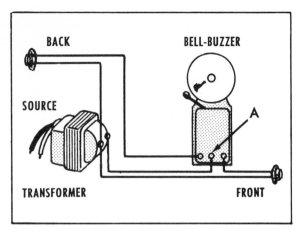

Wiring for a front doorbell and a back-door buzzer.

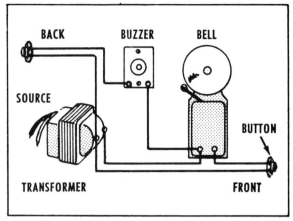

Wiring for a combination bell–buzzer unit.

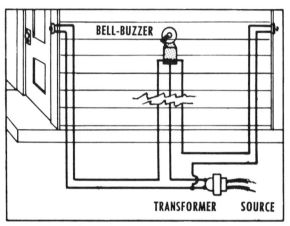

Installed view of a combination bell–buzzer.

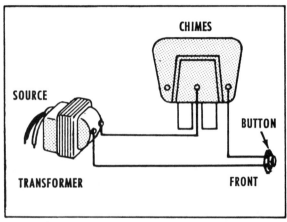

Wiring for a front installation of one- and two-note chimes.

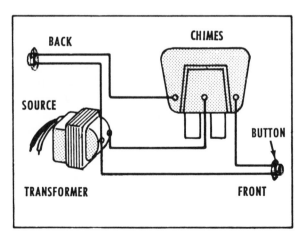

Wiring when a pushbutton for back door is added to the front installation.

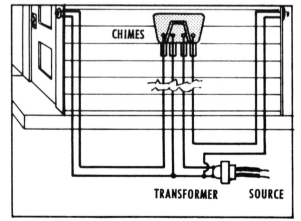

Installed view of a four-note chime setup.

6
Fixes for Plugs, Cords, and Lamps

Electrical breakdowns occur most frequently in lamps or appliances. And in the greatest number of cases, these breakdowns are caused by failure of the lamp's or appliance's plug or cord.

You need not fear to work on a plug or cord. Once it is pulled from the outlet, there is no live current to worry about; you cannot receive a shock. The same is true of any other part of a lamp or appliance. After its cord is unplugged, you can work in safety. **But don't forget! The first step in any repair is to pull the plug.**

A plug has two or more prongs connected by internal terminals to the cord. A cord (also called a line cord) conducts the current from the plug to the lamp or appliance and back, hence, always contains at least two wires. Each is wrapped in its own insulation. The wires are bound together into a cord by this insulation or by a separate sheathing of insulation.

Plug Repair and Replacement

Plugs are available in a bewildering selection of sizes and types. Some are short and squat, some are narrow, and some are long and deep; they come in round, rectangular, and other shapes. Some are *sealed;* that is, the terminals are molded right into them and cannot be reached. Others are of so-called *open construction.* The latter are ideal for general replacement purposes because the terminals are readily accessible. Here is what to do about common plug defects:

Plug Keeps Falling from Outlet: Usually, a plug falls out of an outlet because the prongs are bent or misshapen and cannot grip. Try bending the prongs outward, that is, away from each other. Stick the plug into the outlet. If the prongs still do not grip, bend them out a little more. If the plug still falls out, replace it. If replacing the plug does not solve the problem, something is wrong with the outlet.

Plug Makes No Electrical Contact with Outlet: Here again, try bending the prongs outward a bit. If they are dirty or corroded, use sandpaper or a nailfile to clean them. Some prongs consist of a metal ribbon bent double to form a spring. Try inserting a screwdriver between the leaves of the springs to spread them a little.

How to Replace a Plug

1. *Worn cords are dangerous and may cause burns and shock. Fix this kind of break by cutting off the plug (its molded-in terminals are inaccessible).*
2. *Cut off damaged cord several inches above the break.*
3. *The best replacement is an open-construction plug. After baring wire ends for about ¾ of an inch, remove fiber insulating disk from plug.*
4. *Insert cord through inlet of plug. It is recommended that the wires now be tied in an Underwriters' knot (see drawing).*
5. *Make loop in end of each wire. (If you are using Underwriters' knot, it's better to postpone baring of the wires until this point.)*
6. *Wrap each wire around prong; then push loop under screw terminal. Tighten terminals.*
7. *Snap fiber disk over prongs — and your replacement is complete.*

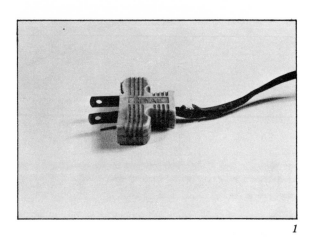

1

2

3

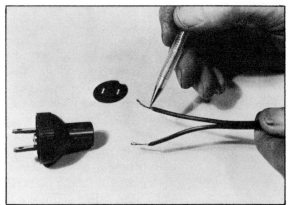

4

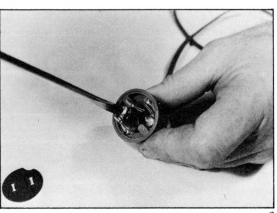

5

6

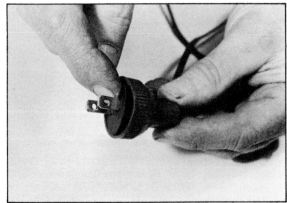

7

Replacement with Snap-On Plug

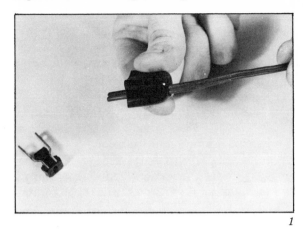

1

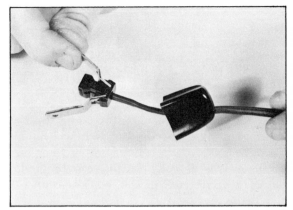

2

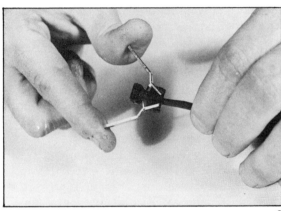

3

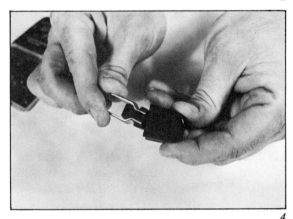

4

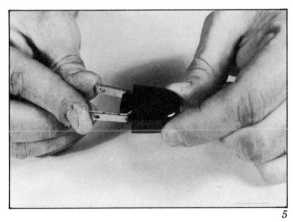

5

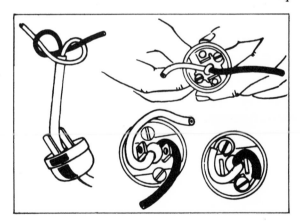

1. *If cord is of molded parallel type, a "snap" plug can be installed — requiring neither the baring of wires nor the tightening of screws. Such plugs vary in manner of installation. To use the kind shown here, begin by slipping the plug body over the cord.*

2. *Spread the prongs and slip the cord into the plug core.*

3. *Make sure the cord is pushed all the way into the core hole; then squeeze prongs together. This causes internal contacts to pierce insulation and bite into the copper.*

4. *Holding the prongs tight against the core, slide the plug body over cord to meet the core assembly.*

5. *Snap the core into the plug body. Plug is now ready for use.*

How to tie the Underwriters' knot. It prevents a break at the plug terminals if you should happen to pull on the cord.

If contact still is not restored, suspect the outlet. Its contacts may be worn, dirty, or covered by paint. Replace the outlet (see Chapter 2).

Broken Plug: A cracked or chipped plug must be replaced — or else sooner or later someone will get shocked. Or a bad short may develop, sending sparks flying.

The best replacement is an open-construction plug, one long enough for a hand to grip it easily. Otherwise, you will be tempted to pull out the plug by hauling on the cord, leading to wire breaks as well as to disconnections at the plug terminals.

Cut off the old plug where it joins the cord. With a knife or razor blade, cut away about 2 inches of any outer fabric or other insulation,

Quick Way to Splice a Cord

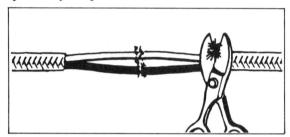

1. Cut away any damaged wire and snip off surplus insulation.

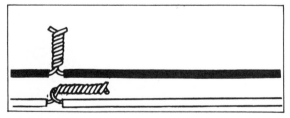

2. Bare the wire ends. Link each wire by twisting ends tight. Bend the twisted connections flat against wires.

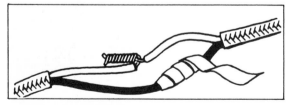

3. Cover each wire with plastic electrical tape, extending tape over insulation. Then bind separate wires together with an overall wrapping of tape.

thus exposing the wires. Bare the wire ends for about 3/4 of an inch. You will find that each wire consists of fine copper strands. Twist the strands clockwise with your fingers to wind them together as tightly as possible.

Approved open-construction plugs have a fiber insulating disk seated at the prong bases. Snap out the disk. Slip the cord into the plug. Tie an "Underwriters' knot"; then, thread one wire behind a prong and to the screw opposite the prong. Bend the bare end of the wire clockwise under the screw. Tighten the screw. Thread the other wire around the other screw; tighten.

Inspect the plug to make sure no fine copper strands have worked loose. If you find any, cut them off with scissors. Push the wires into the plug as deeply as possible. Finally, snap the fiber disk over the prongs.

Parallel-type lamp cord is easily recognizable because the two wires are bound lengthwise by their own insulation rather than by a covering that gives the appearance of a single cord. Special snap-on plugs are made that accept parallel cord without cutting away any insulation or turning any screws. Simply poke the cord into the plug and press the prongs or a small catch, and your replacement is ready.

A snap-on plug will not accept larger sizes of cord and has no great strength. The open-construction type, although less convenient, remains the strongest, most versatile, and safest plug.

Female Appliance Plugs: Some "plugs" are not really plugs at all. Instead, they are receptacles. They have female openings to fit over the male prongs permanently fabricated into certain appliances.

You cannot repair some female plugs since the halves of the casing are held together by rivets. If the plug cracks, chips or fails, your only choice is to buy an entire replacement line cord that comes with a factory-affixed plug of the right kind. But if the halves of the defective plug are held together by screws, you may be able to fix it.

Split the plug by taking out the screws (the halves of some female plugs can be merely snapped apart). Inside you will see two wires attached to terminal screws, as in a male plug. Loosen the screws and clean away dirt and corrosion. Tighten the terminal screws securely over the wires. If the metal lining the two receptacle holes appears distorted, try adjusting it with a screwdriver to assure good contact with the appliance prongs. If the plug still does not work, it must be replaced.

Three-prong Plugs: Many appliances utilize plugs with a grounding prong in addition to the two usual prongs. Heavy-duty 3-prong plugs may be encased in a metal strap that must be removed before the plug can be worked on. In other respects, they are handled the same way as 2-prong plugs. When you dismantle the plug, note whether the wires are colored or threaded to indicate which wire should go to which prong. If you must make fresh connections, follow the directions given above. The three wires should be tied together in a knot resembling the Underwriters' knot before being connected to the screw terminals.

Frayed Cords: Any cord with frayed or damaged insulation should be replaced. Electrical codes frown on the use of a repaired cord. However, if the fray is around the plug at either end of the cord, you can simply cut off the frayed

section and install the plug on the sound section. That makes the cord as good as new — and it satisfies even the most stringent codes.

If the fray occurs somewhere toward the middle of the cord, you might wish to repair it for temporary or emergency use. Use scissors, snips, or cutting pliers to cut the frayed area right out of the cord. You now have two shorter cords, each containing two or three wires. Bare the ends of the wires. Splice each wire of one cord to a wire of the other. Wrap each splice in electrical tape, overlapping for at least an inch the insulation at both ends of the splices. Then wrap the bundle in electrical or friction tape.

Before taping, it is better to solder the splices. For splicing and soldering directions, see Chapter 4. When soldering cord splices, it is perfectly safe to use an electric soldering iron. The fact that the cord is disconnected from any outlet eliminates all possibility of shock.

Fixing Lamps

Here's another job well within the grasp of even the most inexperienced. You don't have to be told that should a lamp flicker or fail to light the first thing to do is change the bulb. **If that**

Replacing a Lamp Cord

1. *After disconnecting lamp from outlet and removing the shade or reflector, unscrew the bulb.*

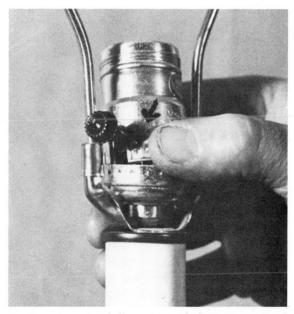

2. *Press the socket shell at point marked* press *(arrow) and lift free.*

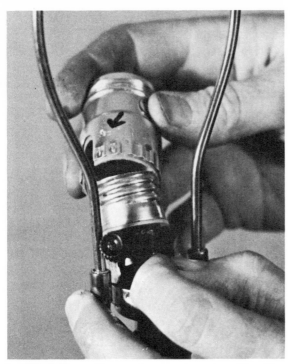

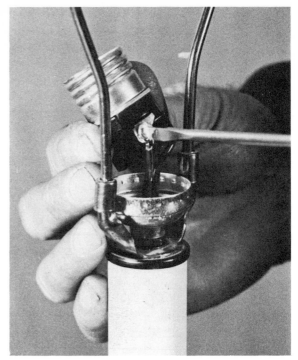

3. *Lift shell out of base and remove the casing. (Note fiber liner peeking out of top and bottom of casing.) Switch is exposed.*

4. *To release cord, loosen terminals. (At this stage you can replace switch and entire socket if they are defective.) Install new cord; reassemble the lamp.*

does not help, pull the lamp's line cord from its wall outlet to eliminate all shock hazard. Then dismantle the lamp.

Start by removing the shade and bulb. On the socket shell you will see the word *press*. Place your thumb on the word and do just that — press — at the same time vigorously twisting the shell and pulling it upward. It should break loose from the socket base. If it won't, pry it out with a screwdriver inserted between shell and base at the press point. This will distort the shell and base, but later you can bend them back to approximately the correct shape.

Inside the shell is an insulating fiber liner. It may come off with the shell. If not, lift the liner from the base. The two screw terminals on the socket base holding the two wires of the line cord are naked now. There you may see dirt, corrosion, or a loose — perhaps broken — wire. Sand the metal surfaces clean. If necessary, cut wire ends, bare off old wires for ¾ of an inch, and connect them anew to the terminals.

Possibly now, upon reassembly, the lamp will work. If you are not that lucky, tackle the switch.

5. *Tie a simple knot at base of lamp stem to keep the section of cord within the lamp fairly taut. This prevents cord abrasion and loosening of terminals.*

Feed-through switch can be easily installed at any point along cord — so that lamp can be controlled, for example, from bedside.

It may be of the pull-chain, push type, or the turn type. If the latter, turn the switch button counterclockwise and it will come off. Feed some slack into the line cord by pushing it up from under the base of the lamp. Twist the socket base counterclockwise; this unscrews it from the lamp. At

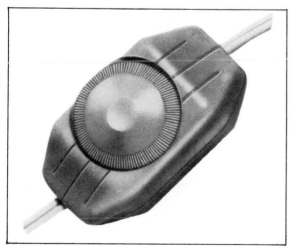

Feed-through switch can be of modern dimmer type, permitting full range of brightness.

a store, buy a replacement socket of the kind that contains a switch. Put the lamp back together by reversing the procedure you used to dismantle it.

In some kinds of table lamps and especially in floor lamps, a screw goes through the socket base to clamp it to the lamp. You will not be able to remove the socket unless you first loosen that screw.

If the lamp still will not work, there is a final possibility. There may be a break in the line cord. You can check for such a break by using your low-voltage tester. Touch one probe to one end of a wire and the other probe to the other end of the wire. Do the same for the other wire. Both times, the tester bulb should light. In any event, if you have reached this final step, changing the cord is quite easy in table lamps. Unscrew the socket terminals to release the wire ends. From the bottom of the lamp, pull the wires right out. Drop your new cord down through the hollow lamp core. Grab it at the bottom and pull it through.

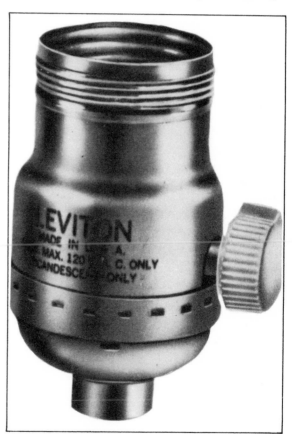

Replacement lamp sockets now are also available with the dimmer feature.

Then bare the top ends, connect them to the terminals. Reassemble the lamp.

Floor lamps may be more difficult to thread with the new cord because the distance to be traveled is greater. Sometimes it is necessary to dismantle the lamp body itself into shorter sections through which the cord can be pulled. Lamp sections, whether within an outer shell or not, are held together by hollow couplings, studs, and other devices, including a large "wheel" under the base to give it weight so that the lamp will not topple. All these fittings are threaded. By turning them counterclockwise or clockwise, you can take the sections apart or lock them together.

7
Repairing Appliances

The modern home is dependent on electrical appliances for convenience and comfort. When these appliances break down, their loss is keenly felt. But often, the do-it-yourselfer can repair them easily.

If your $500 electric range refuses to work properly or at all, it's probably best to call in an expert to get it back in shape. But if a toaster or some other small appliance malfunctions, chances are you can fix it yourself. **But remember — to avoid shock, work on an appliance only when its line cord is not plugged into an outlet.**

Before You Begin

Make sure that you carefully read its accompanying instructions before using any appliance. And if you still have those instructions around when an appliance stops working, dig them out and see if they might help. Have you, perhaps through carelessness, tried to get the appliance to do something it wasn't designed to do? Was it overloaded? Did someone use it who was unfamiliar with it — a child, a visitor, the cleaning woman?

External Supply Failure: Is the cord's plug making contact in the outlet? Check the fuses, too.

Physical Damage: If an appliance has been through a fire or flood, or even if it was dropped recently, this obviously may present problems. If fire was the culprit, the problem is probably taken care of by your insurance policy. But flooding is rarely covered by insurance. So clean the machine thoroughly and bake the electrical parts dry in a low oven (180°). Check all contacts for cleanliness and replace damaged insulation by winding it with plastic electrical tape. Make sure all connections are tight. Lubricate the moving parts and remove all traces of rust. If the parts still don't operate smoothly, replace them. Usually, you can order replacement parts from the dealer or manufacturer.

If an appliance has been dropped, check the connections. See that everything lines up properly. If the plastic covering was broken by the fall, make sure that tiny particles do not remain in the moving parts. Correct dents in metal surfaces that may be pressing against moving parts inside.

Operational Failure: If neither external supply failure nor physical damage is responsible for failure of the appliance, you have to do some diagnosing. Stop, look, listen, and — most important — use your nose. If, when any electrical appliance is plugged into a receptacle, you smell burning insulation, the machine should be disconnected promptly. Remove the screws holding the cover in place, remove the cover, and look for wires with burned-out or smudged insulation. Reinsulate with electrical tape or replace with fresh wire. If insulation is burned off a wire, all of the parts connected by it should be checked for "grounds" or "shorts" with a line tester (see below). If a part looks carbonized, eroded, or otherwise questionable, it is safer and less expensive in the long run to replace it.

Noises Are Giveaways: A rapid rattling noise when the appliance is plugged in may mean that a foreign object, such as a hairpin or a marble, has found its way inside. A "loose," more erratic noise may just require tightening a few screws.

Occasionally, from motor-driven appliances, you hear a grinding noise. This indicates a bad bearing. Fixing a bearing is generally a special-tool job that calls for outside help.

Don't forget proper and regular lubrication of motors if the manufacturer's instructions call for it. A squeaky noise may mean you have neglected this. Slipping belts make characteristic sounds: slaps and squeaks. Most belt-driven motors have mechanisms that permit the easy adjustment of belts. If not, replacing a belt is not costly.

Where open lids or other parts cause an automatic shut-off, see that they are properly attached and closed, with contacts meeting. If the contacts are dirty, clean them with sandpaper or a nail file. Keep dust and dirt to a minimum.

The Underwriters' Knot: Nine times out of ten, the trouble with an appliance lies in its line cord or plug. So a basic simple chore in appliance repair is plug replacement. Using the Underwriters' knot (see page 61) when you replace a plug assures you that everyday strains will not cause the cord to pull loose from the plug. Use this knot when replacing a plug, but make certain that the cord leading from the plug to the appliance is otherwise without kinks or knots.

Testing: Tests are basic to appliance repair. Testers of various types are available in most hardware and electrical supply stores. The 120-volt tester (see Chapter 1) will not do for testing an appliance. What is needed is a low-voltage tester containing its own current source — a small battery. With minor variations, such devices all work the same way. They consist essentially of a miniature low-wattage bulb connected to the battery and two leads. The leads are touched to the bare ends of a length of wire or to the terminals of a part, such as a heating element. If the bulb fails to light, you know there is a break in the wire or heating element. If so, replace the wire or element.

Electric Motors

Most small appliances have either a motor or heating element; some have both. You should know if the motored appliance you are repairing has a "brush-type" or a "synchronous" motor.

Brush-type Motors: These motors require maintenance — mostly an occasional oiling (if specified by the manufacturer) but also some attention to the brushes. Instead of the simple rotor found in a synchronous motor (see below) the brush-type motor has a more complicated version of the same thing. This is called an "armature." The armature revolves within a surrounding electromagnet. The electricity that activates both the armature and the electromagnet passes from one to the other by means of "brushes." These maintain contact between the spinning armature and the fixed electromagnet.

It is important for you to realize that brushes wear out. They are made of soft carbon, and in the course of time, they must be replaced. This is a simple task, involving only the loosening of the two large-headed screws on either side of the motor housing. After they are unscrewed, the worn brushes will pop out on springs. Be sure to replace the brushes with ones of exactly the same size and cross-section. You can buy replacement brushes in any electrical supply store. Occasionally, sandpapering the copper bars of the armature will restore efficiency.

These bars, known collectively as the "commutator," must maintain good contact with the brushes.

Synchronous Motors: These motors are smaller and simpler. They lack brushes and require little or no maintenance. Hence, they have become more and more popular in recent years for use in small appliances. They consist basically of two elements — an electromagnet and a rotor. The magnet consists of a U-shaped laminated frame around which many turns of fine wire are wrapped. The wire is plugged into a power source, creating a magnetic field. In an AC circuit, this field changes direction 120 times a second, causing the rotor to revolve. These motors are excellent where a constant (synchronous) speed is desired. They are used in electric clocks, record-player turntables, and other constant-speed appliances.

Toasters

Your toaster is a simple mechanism. It consists of line cord, heating elements in a housing, a timer, and/or a thermostat. Typically, an automatic toaster has two heating elements — the middle and one on the outside — between which each bread slice is sandwiched.

Many toaster problems are caused by crumbs collecting in the mechanism. A thorough cleaning every few months is advisable. Most toasters have an easily removable plate on the bottom for crumb-cleanout. Actually, there is not much to repair in the automatics. Rarely someone pokes around with a fork and ruins a heating element (or gets killed by the jolt). Replacement elements can be bought from a store or the manufacturer. At times, the thermostat or timer may be balky. In such cases, it is easier and cheaper to buy new parts than to repair the old. Also, line cords and plugs go wrong, but they are easy to replace.

Mixers

A common problem with mixers is failure of the beaters to turn. This may be caused by two things: either the beaters are bent or defective or the beaters' gears are worn. Visual inspection should tell you whether or not the beaters themselves are okay. If not, they are easily replaced. Handling the gears is a little more complicated. First, carefully scrutinize the mechanism to determine how to take it apart so that the gears may be inspected. You will probably find that the gears are stripped, but it may be just a case of plain wear and tear.

When replacing defective gears, make sure

Troubleshooting Chart: Toasters

Problem	Cause	Solution
No heat	No power at outlet	Check outlet, fuse
	Defective cord	Repair or replace
	Loose connection	Clean, tighten
	Switch not making contact	Repair or replace
	Elements burned out	Replace
Toast will not stay down	Hold-down latch not locking	If bent, straighten; if binding, clear to allow free operation
	Bind in toast carriage	Clear cause of bind
	Broken latch spring	Replace
	Timer binds	Repair or replace
Toast will not pop up	Bind in toast carriage	Clear cause of bind
	Release latch binds	Clear cause of bind
	Broken spring	Replace
	Timer binds	Repair or replace
Toast too light/dark	Thermostat or timer malfunction	Adjust or replace

Troubleshooting Chart: Mixers

Problem	Cause	Solution
Doesn't work at all	No power at outlet	Check outlet, fuse
	Defective cord	Repair or replace
	Worn brushes	Replace
	Broken field coil	Replace
	Broken armature winding	Replace armature
	Defective switch	Replace
Doesn't run, blows fuses	Bent shaft jamming armature	Straighten or replace shaft
	Defective armature or field coil	Replace
	Shorted cord	Repair or replace
Motor runs hot	Bind in shaft	Clear bind
	Shorted winding in armature	Replace armature
	Shorted field coil	Replace
Motor runs, beaters don't turn	Stripped gears	Replace
Erratic operation, speed	Worn brushes	Replace
	Loose connection	Clean and tighten
	Defective switch	Replace
Slow speed, weak power	Incorrect setting	Adjust speed control, reset
	Worn brushes	Replace
	Bind in shaft	Clear bind
Too noisy	Armature hitting field	Replace worn bearing
	Bent cooling fan blade	Straighten
	Dry gears or bearing	Lubricate

that you position the new ones so that the beaters do not strike each other. The gears should be at a 45-degree angle to one another to insure free beater rotation. Test the machine with the beaters in place before putting it back together. If they touch, try again until they are completely independent of each other. In mixers with brush motors, the brushes also should be replaced in pairs. Such mixers usually have brush caps on the outside of the motor housing so that inspection is easy. If the brushes are worn to less than ⅛ of an inch from the spring, they should be replaced. Check the springs, too, to see if they are damaged or weak (see troubleshooting chart).

If the brushes are not smooth, clean, and curved to the surface of the commutator, you may find that the armature turns erratically. If the trouble is in the armature itself, it may be wisest to get a new mixer.

Incidentally, you can forestall trouble by giving your new mixer a break-in period. Run it at lower-than-normal speeds and on not-too-thick mixes for a few days until it is ready to flex its muscles.

Coffee Makers

The problems that usually plague electric coffee makers are defective water heating and build-up of flavor-killing deposits within the pot.

Defective water heating results in coffee that is too weak (flavor left in the beans) or too hot (bitter oils extracted along with the flavor). Water should be between 175° and 190° when it passes over the coffee. If it isn't, check the troubleshooting chart below. Deposits can be built up in the pot by hard water or as a result of improper cleaning. Hard water will leave lime deposits, and improper cleaning will leave coffee solids containing bitter oils. A strong vinegar solution will dissolve any lime deposits. A regular schedule of "boiling out" the pot with baking soda will help eliminate deposited coffee solids.

The perking action of a coffee maker is not caused by boiling of the water, as is commonly believed; instead, it is caused by steam. Steam is generated under the basket-stem base in the heater well. The steam "pumps" water up through the percolator tube in spurts; the water then splashes down over the coffee grounds, extracting their flavor. The flow is controlled by the pump check valve, a loose disk at the bottom of the stem. As a charge of water is perked up, more water flows into the well through the valve, where in turn it is pumped up to the top of the tube and out. It's important on all pump-type coffee percolators to keep the disk and valve-seat clean, smooth, and unscratched. It is equally important to keep the stem clean to prevent clogging.

Actually, most coffee makers (other than drip types) have two heaters. One starts the perking action, while the other acts more slowly to heat up the water. When the water is sufficiently heated, a thermostat maintains the temperature until the line plug is pulled from its receptacle. Up to that time, the thermostat remains closed so that the full line voltage can reach the pump heater.

A good coffee maker is worth repairing. Most parts that might need replacing (such as pump heaters, cords, and thermostats) can be purchased at reasonable prices.

Rotisserie Oven-Broilers

A typical broiler consists of a thermostatically controlled heater element, a motor to operate the spit, and a timer to stop the cooking after a preset time.

Spills from overcooking are a common cause of trouble. If this happens, it is much easier to clean up the mess right away than to bake on the deposits through repeated usage.

Make sure the oven is cool and unplugged before you wash it. The glass, particularly, may break if it is wiped or washed while still warm. Dry by hand; then, turn on the oven to 250° and let the appliance run for a while. This should prevent rusting.

Electric Frying Pans

Electric frying pans are simple appliances. They contain no moving parts — just a heating

---

Troubleshooting Chart: Coffee Makers

Problem	Cause	Solution
Doesn't operate	No voltage at outlet	Check outlet, fuse
	Defective cord	Repair or replace
	Defective pump heater element	Replace
Gets warm but doesn't percolate	Defective pump	Replace
	Defective thermostat	Replace
	Incorrect setting of thermostat	Reset
Slow in brewing coffee	Low line voltage	Call power company
Coffee tastes bitter	Accumulated residue inside maker	Clean with baking soda or other cleaner
Weak coffee	Control incorrect	Reset
	Using hot water to start	Use cold water
	Pump valve stuck	Clean valve or replace if damaged
Coffee boils	Incorrect thermostat setting	Adjust
	Defective thermostat	Replace
	Valve stuck; stem clogged	Clear blockage in stem

Troubleshooting Chart: Rotisserie Oven-Broilers		
Problem	*Cause*	*Solution*
Won't operate	No power at outlet	Check outlet, fuse
	Defective cord	Repair or replace
	Defective switch	Replace
	Defective timer switch	Replace
Heats but motor doesn't run	Defective motor	Replace
	Stuck gearing	Clean gears
	Defective motor switch	Replace
Motor runs but oven doesn't heat	Defective heater element	Replace
	Defective heater switch	Replace
Incorrect heat	Defective thermostat	Replace

element and a cord. There is little difference among brands except that some are immersible, and others are not (a distinction printed clearly on the nameplates).

If a nonimmersible frying pan is inadvertently submerged in water, take off the bottom plate as soon as possible. This can be done by removing a few screws. Dry off the unit with a fan or vacuum cleaner blower — or better still, use a blower-type hair dryer if you have one.

Electric pans do not go wrong easily. Other than unsanctioned immersion, problems are usually caused by a malfunction in the temperature control switch or the line cord.

These pans cook faster at lower temperatures than do your ordinary nonelectric pans, so the habit-ridden homemaker often turns the heat too high for too long. The result is sticking, burning, and a difficult clean-up job. The best way to clean an electric pan is to boil a couple of cups of water in the pan for a few minutes; then unplug the unit and pour off the water. Scrape out the residue with a wooden spoon or spatula; then wash out the inside with hot water and a mild soap. Strong detergents, alkaline cleaners, and abrasives should never be used. Electric frying pans have a surface that must be "seasoned." That is, the porous surface must be sealed with fresh shortening after every vigorous cleaning.

Although a sturdy appliance, the electric frying pan is more susceptible to external damage than some other appliances. Frequently a hard blow will change the thermostat's calibration.

A thermocouple is the best tool for checking and resetting it. But you can attain a reasonably close approximation by noting the setting at which water boils. If it is within 20 degrees, either way, of 212°, the calibration is close enough — unless you live on a mountain or in a mine.

When reassembling an immersible frying pan after making any repairs, take care to reseat all gaskets properly. If a gasket looks worn, or if it won't seat properly, replace it with a new one. The cost is negligible.

Waffle Irons

A waffle iron is electrically simple; it consists of a resistor connected to a plug. Some models have a thermostat and removable grids so that they can be converted into a grill. The thermostat tells you when the grids are hot enough to use and keeps the inside temperature constant. The resistor — a coil of nichrome or similar wire — serves as a heating element.

The entire electrical system is easy to get at in the newer models. Typically, the grids lift right out and you see the long coiled heaters, a light and a thermostat. When the iron is hot enough, the thermostat breaks the circuit, shutting off the light. Once the light is out the appliance is ready to use — unless there is a problem.

Visual inspection should reveal any breaks in the coil. A broken coil should be replaced

Troubleshooting Chart: Electric Frying Pans

Problem	Cause	Solution
No heat	No power at outlet	Check outlet, fuse
	Defective cord	Repair or replace
	Broken heater element	Replace
	Defective thermostat	Replace
	Poor connection	Clean and tighten
No heat control	Defective thermostat	Replace
Shocks user	Grounded unit	Replace defective part
	Wire touching frame	Locate and reinsulate or replace wire
Food sticks	Excessive cooking temperature	Lower thermostat setting
	Pan is not seasoned	Season pan; heat for half hour with shortening

unless the break is within an inch of the end post; you can stretch it that far and reconnect it. Be sure to tighten screws all the way; never allow looseness in the coil. Direct contact of the coil with the plates could cause a fatal accident. As a matter of fact, after any such reassembly, the iron should be checked with your tester for a short. With the unit unplugged, place one lead on a power terminal and the other on the shell to see if any current is escaping. The old, usually round, waffle irons are wired in parallel, so it is easy to determine if a coil is defective. If one side works and the other side won't,

it will be necessary to replace the latter.

Waffle grids can be black or shiny or in almost any color or condition, except unoiled. In other words, they must always be "seasoned" (coated by an application of cooking oil or shortening). If a grid has been burned too badly, it requires patience and hard work to remove all the batter. But it must be removed. Don't use a detergent unless absolutely necessary. If you must use one, the oil will be drawn out of the pores. The iron must then be seasoned again, just like a new one. As an extra precaution, add a bit more cooking oil to the batter.

Troubleshooting Chart: Waffle Irons

Problem	Cause	Solution
No heat	No power at outlet	Check outlet, fuse
	Defective cord	Repair or replace
	Damaged heater element	Replace
	Broken hinge wire	Replace
	Defective thermostat	Replace
Blows fuse	Shorted cord	Replace
	Shorted heater element	Replace
	Shorted wiring	Reinsulate wiring
Too hot	Check thermostat setting (maximum temperature: 520°)	Reset or replace thermostat
Waffles stick	Improperly seasoned grid	Operate for half-hour with cooking oil — no batter — on grids
	Insufficient shortening in batter	Add more cooking oil
	Too much sugar in batter	Correct excess
	Opening griddle too soon	Patience!

Troubleshooting Chart: Blenders

Problem	Cause	Solution
Motor won't run	Problem at outlet	Check fuse, outlet wiring
	Defective line cord	Repair or replace
	Defective switch	Repair or replace
	Burned-out motor	Replace armature or field coil
	Frozen bearings	Free, lubricate
	Armature hitting because of worn bearing	Replace bearing
Motor runs, blade doesn't turn	Broken belt (on some models)	Replace belt
	Incorrect placement of container	Relocate on base
	Defective motor coupling	Replace
Runs at high speed only	Defective switch	Replace
	Open resistor	Replace
	Defective field coil	Replace
Blade damaged	Hitting ice cubes, spoons, bones, etc.	Replace
Container leaks	Cracked glass jar	Replace
	Poor seal	Tighten bushing or replace seal

Blenders

These handy gadgets will chop, blend, grind, shred, or liquefy foodstuffs fed to them. But a blender is not an ice-crusher. To use ice in most blenders, you have to crush the cubes first — or the blades may suffer.

Other than dropping and breaking the blender, problems are caused most often by damage to the cutting blades. These will chip or curl if nonblendable objects, such as whole ice cubes, are fed to them. Blades are changed by holding the knife shaft with a wrench and turning the spiral retaining nut clockwise. Next, remove the blades. Take off the nut and bushing; then insert new blades and reassemble the blender.

Irons (Dry or Steam)

A standard electric iron is nothing more than a resistor taking current from the line with a thermostat controlling the temperature.

The dial that indicates "wool," "rayon," etc., merely causes the thermostat to allow more or less electricity to get to the heating element. It is important, when disassembling an iron, to mark where the indicators point. Otherwise it may be difficult to match them up upon reassembly.

To check the setting of a thermostat, use a thermocouple or a high-range thermometer like those used for deep-frying. Set the iron on some heatproof material, such as asbestos or fiberglass. Place the bulb of the thermometer under the iron about a third of the way back. Turn the control to a few settings and verify with the thermometer. If temperatures are not given on the control, figure 225° for rayon, 350° for wool, and 525° for linen. Variations can usually be corrected by turning the small setscrew in the center of the control shaft.

Dry and steam irons share a common fault. The line cord frequently breaks down, particularly at the cord sleeve next to the iron. In such a case, cut off the cord close to the plug, bare the two leads, and refasten in the plug. Always wrap thread around the cut end of a heater cord to keep the loose asbestos filler from falling out. Breaks elsewhere in the cord may be suspected where the exterior looks frayed, burned, or is soft and bends too easily. The best practice is to buy a new cord rather than attempt to mend it.

Steam irons have a special problem with the hard water found in many parts of the country. The minerals in the hard water accumulate and clog the small steam portholes and passages. This problem can be prevented by using a water softener or distilled water. Rain water, melted snow, and defrosted water from accumulated ice from the refrigerator are good sources of mineral-free water.

· If it is too late for any of these preventives, the iron can be fixed by filling the tank with vinegar — which will dissolve the lime deposits. Heating the iron with the vinegar inside may be necessary, and the process may have to be repeated.

Troubleshooting Chart: Irons

Problem	Cause	Solution
No heat	No power at outlet	Check outlet, fuse
	Defective cord, plug	Repair or replace
	Broken lead in iron	Repair or replace
	Loose connection	Clean and tighten
	Loose thermostat control knob	Replace knob and tighten
	Defective thermostat	Replace
	Defective heater element	Replace heater if separate; replace soleplate if cast in
Insufficient heat	Low line voltage	Check voltage at outlet
	Incorrect thermostat setting	Adjust thermostat
	Defective thermostat	Replace
	Loose connection	Clean and tighten
Excessive heat	Incorrect thermostat setting	Adjust thermostat
	Defective thermostat	Replace
Blisters on soleplate	Excessive heat	Correct condition (above); repair or replace soleplate
Water leakage	Defective seam or tank weld	Replace tank
	Inadequate tank sealer	Reseal with proper sealer
	Damaged gasket	Replace gasket
No steam	Thermostat set too low	Set control higher
	Valve in off position	Turn to correct position
	Dirty or plugged valves or holes	Clean out
Spitting	Incorrect thermostat setting	Reset thermostat higher
	Excessive mineral deposit	Clean out
	Overfilling	Drain, be more careful
Bad spray (spray irons)	Defective plunger	Replace
Stains on clothes	Starch on soleplate	Rub soleplate with damp cloth, polish with dry cloth
	Foreign matter in water	Use distilled water
	Sediment in tank	Clean with vinegar
Tears clothes	Rough spot, nick, scratch or burr on soleplate	Remove with fine emery, then buff or polish
Sticks to clothes	Dirty soleplate	Clean
	Excessive starch in clothes	Iron at a lower temperature; use less starch

Troubleshooting Chart: Fans — Window or Portable

Problem	Cause	Solution
Won't run; fan can be turned by hand	No power	Check outlet connection, check fuse
	Break in cord	Repair or replace
	Defective switch	Repair or replace
Fan won't turn	Blades hitting grille	Straighten grille, blades
	Armature hitting starter in motor	Usually indicates worn bearings, rust or foreign matter; clean, replace worn bearings
	Armature frozen, needs lubrication	Free armature, lubricate
	Misalignment of bearings	Realign bearings
Runs slow	Bearings dry, gummy, or misaligned	Clean, lubricate, realign bearings
	Defective speed control	Replace
Noisy	Fan blades hitting	Clear obstruction, straighten blades
	Fan blades bent	Straighten blades
	Worn bearings	Replace bearings; where this is not practical, replace motor

Fans — Window or Portable

Basically, an electric fan is not much more than a motor with blades attached to the shaft. It is a simple mechanism, and the problems likely to be encountered are simple also. Don't complicate the job by taking apart more than you have to.

To disassemble a fan, first remove the guard. Often a fan has a snap-on guard, which is removed by carefully lifting it over the other sections. If there are screws, remove them. The fan-blade assembly is attached to the motor shaft by a setscrew, which should be loosened. The motor housing has two or more sections held together by two or four bolts. Most fan motors are cooled by the blade motion so there are holes in the motor housing that allow the air to enter and circulate. This, of course, also brings dust into the motor, so an occasional cleaning is in order.

Possibly the most common complaint about fans is noisy operation. This is usually caused by unbalanced fan blades (see troubleshooting chart above for other possible causes). The easiest way to align unbalanced blades is to remove the entire assembly and place it, blades down, on a workbench or other flat surface. Make sure the shaft bore is absolutely perpendicular, and then check to see if all blades touch the surface. If there is only a slight misalignment, bend the offending blade by hand or with pliers until it matches the others exactly. Use a vise to put severely bent blades back into shape.

If the blade appears perfectly straight and if there is no looseness in the oscillator mechanism (the part that makes some fans move right and left), the trouble could well be the motor bearings. The clue is excessive movement on the shaft when the blade is wriggled by hand. There should be only a slight amount of play. If worn bearings are indicated, take the motor completely apart and replace with fresh bearings. In less expensive fans, however, bearing replacement may be virtually impossible. You will have to buy a new fan or at least a new motor.

Portable Electric Heaters

Few homes have such a perfect central heating system that a portable heater does not come in handy. A heater is one of the simplest mechanisms known. It is composed of a heating element, a protective grille, and a reflector to radiate the heat. Thermostat control is standard on most units. This is set for a certain room temperature, and the heater turns on whenever the temperature falls below it. Many units also include a fan to circulate the warmed air.

If the heater does not shut on and off when you think it should, the thermostat may be inaccurate. To adjust it, remove the knob and adjust the setscrew in the center of the shaft. If the heater will not shut off, unplug it and turn the knob to the *off* position. Clamp the leads of your low-voltage tester to the plug contacts and keep turning the setscrew in a counterclockwise direction until the tester light goes off. If turning the setscrew makes little difference, a new thermostat may be needed.

Most heaters have a removable back that exposes the wiring for point-to-point checkout with the tester. While you're taking the heater apart, it is wise to clean out any dust. A vacuum cleaner and a small paint brush will do an excellent job. It is important to clean the reflector regularly. In addition to dry wiping, a periodic cleaning with mild soap and a wet sponge will increase the heater's efficiency.

The heating element can be visually inspected after removing the grille. A break in the heating element is easily observed. Do *not* shorten the resistance wire and connect to the nearest terminal if you don't want to blow a fuse. You can save the element temporarily by silver-soldering it — or you can buy and install a new element.

The wiring in an electric heater is almost always asbestos-insulated. If it must be reinsulated, make certain you use either asbestos or glass tape. Any other material is apt to soften, smoke, and smell as the heat climbs.

Fan-equipped units that are excessively noisy in operation can often be quieted simply by bending the blades slightly.

Electric Shavers

If you clean out your shaver every once in a while, lubricate it, and don't drop it, you should have no problems. A shaver housing is easily opened by removing a screw or two. All too often, though, shavers are dropped, resulting in a

Troubleshooting Chart: Portable Electric Heaters

Problem	Cause	Solution
No heat	No power to unit	Check fuse, outlet, wiring
	Defective cord, plug	Repair or replace
	Defective switch	Replace
	Defective thermostat	Replace
	Defective heater element	Replace
Low heat	Incorrect thermostat setting	Adjust thermostat
	Defective heater element	Replace
Won't shut off	Defective switch	Replace
	Defective thermostat	Replace
	Shorted wiring	Locate, separate wires, reinsulate
Fan doesn't run	Loose connection	Locate and tighten
	Jammed fan blade	Straighten
	Frozen motor bearing	Free armature, lubricate bearing
	Burned-out motor	Replace
Excessively noisy	Fan blade hitting obstruction	Straighten blade or clear obstruction from blade
	Worn motor bearings	Replace
Shocks user	Defective wiring	Clear bare wire from frame and insulate
	Heater element touching reflector	Repair or replace element

Troubleshooting Chart: Electric Shavers

Problem	Cause	Solution
Does not run	No power	Check outlet receptacle, fuse, wiring
	Defective cord, plug	Repair or replace
	Switch not making contact	Bend contact arm
	Contact points burned out	Replace
	Shorted capacitor	Replace
	Burned-out motor	Replace
Runs too slowly	Dry motor bearing	Lubricate
	Binding or dry shaving head	Clean, lubricate
	Damaged shaving head	Replace
	Electrical defect	See below
Excessive sparking	Worn brushes	Replace
	Dry bearing	Lubricate
	Improper timing	Adjust timing
	Defective capacitor	Replace
	Defective contacts	Replace
	Defective suppressor resistance	Replace motor winding
	Shorted field winding	Replace motor winding
Shaves poorly	Running too slowly	See above
	Worn shaving head	Replace

damaged case or head. The only smart thing you can do is replace them. Sometimes you can glue or even tape the case together, but you cannot use a shaver with a broken tooth or bar.

The cord, too, gets a lot of wear. If the shaver is not working at all or if it has a tendency to cut out on you, there may be a break or short in the cord. Check it out with your tester. A new cord may be in order.

Vacuum Cleaners

You may not think of a vacuum cleaner as a small appliance, but it is a relatively simple machine (despite its price), and many malfunctions can easily be repaired by the competent do-it-yourselfer.

Although vacuum cleaners are divided into two rather distinct types — upright and tank — their differences are based more on styling and customer preference, than on any real dissimilarity. Essentially, a vacuum cleaner of either type consists of a motor-driven fan that creates suction. The larger upright models boast a revolving brush to "beat" the rug and loosen dirt. This helps the upright do a better job on rugs, but the tank-type is generally conceded to have more versatility and drawing power.

The fan section of the housing is sealed off from the motor area, so that none of the dirt gets into the motor. There is an opening through which the dirt is sucked into a cloth or paper bag. The bag is of such consistency that it allows the air to escape, yet retains the dirt. On most models, the bag is disposable.

One common reason for poor and noisy operation is wear of the brush bearings. These brushes rotate at very high speed, so that dirt is literally forced into the bearings — causing friction. The V-belt often rides off the brush assembly, too, and the brush must be loosened before the belt can be put back on.

When foreign objects such as string or paper clips are picked up, the fan may jam and the machine will lose suction power or fail to run at all. This happens a great deal in the smaller models. The housing must be opened up (usually this is not difficult) to remove the objects. Occasionally, a foreign object will damage a fan blade, causing it to run abnormally and make an unearthly noise. In such cases, the fan blade should be replaced.

Troubleshooting Chart: Vacuum Cleaners

Problem	Cause	Solution
Motor does not run	No power	Check outlet, fuse, wiring
	Defect in cord, plug	Repair or replace
	Defective switch	Replace
	Worn brushes	Replace
	Jammed fan	Free; if bent or damaged, replace
	Frozen bearings	Clean and lubricate; if worn, replace
Motor starts and stops	Intermittent break in cord	Locate and repair or replace
	Loose connection within cleaner	Check all connections; repair
	Defective switch	Replace
	Loose connection in motor	Check motor; tighten connection
Motor runs too slow, no power	Foreign object caught in fan or armature	Remove object
	Misaligned, tight motor bearings	Realign
	Poor brush contact	Correct or replace brushes
Motor runs too fast	Overfilled dust bag	Replace if disposable, otherwise clean
	Fan loose on shaft, not turning	Check fan balance, tighten nut securing fan to shaft
Motor sparks	Dirty commutator (oil or dirt)	Clean with fine sandpaper
	Worn brushes	Replace
	Incorrect brush seating	Correct seating
Motor too noisy	Foreign matter in motor	Clean out
	Fan damaged	Replace
Poor pickup	Worn or damaged attachments	Check attachments for leakage and replace as necessary
	Incorrect nozzle adjustment for carpet nap	Adjust for correct contact
	Leaky hose	Check for air leaks; repair or replace
	Clogged hose	Blow or push out obstruction
	Overfilled dust bag	Replace if disposable; otherwise clean out
	Clogged exhaust port	Clear
Dust leakage	Holes in dust bag	Replace
	Incorrectly installed dust bag	See owner's manual
	Old, dirty dust bag	Replace
	Defective sealing	Replace

Electric Knives

Electric knives consist of two knives working in opposition. Thus, a fast reciprocating action is achieved that makes short work of slicing. The typical electric knife has a two-part plastic housing. Two screws at one end hold the parts together; lugs fit the other ends together. The motor is at the far end, and a shaft runs into a gear box. In the gear box, the rotary action of the motor turns a pinion by means of a worm gear. The blades are attached to opposite sides of the pinion gear. As one knife moves, the other moves in the opposite direction, creating a shearing action.

The things that go wrong with an electric knife are those that go wrong with most motor-driven appliances: brushes wear out, switch contacts become clogged, armatures burn out. In an appliance of this type, the cord becomes worn because it is frequently in motion. The cord usually has a nonstandard plug where it connects to the knife — a plug like those used in television sets. If your knife has a worn cord and you can't get a replacement, try a standard cheater cord from the local television repair shop.

Electric Can Openers

The electric can opener is simply a mechanical opener operated by a motor rather than by hand. Some openers have a knife-sharpener attachment. This may involve extending the rotor shaft and attaching a grinding wheel. The electric opener employs a shaded-pole, or synchronous, motor, and not much can go wrong with the mechanism unless it is abused. The opener should be cleaned thoroughly and lubricated at least twice a year. (Consult the manufacturer's instructions, if you still have them.)

To clean or sharpen the cutter assembly, remove the spring tension screw and the screw in back of it. If you have to check the motor, the housing can usually be removed by loosening screws on the bottom of the opener. This does not disturb the internal workings. But electric can openers, if maintained as described above, are relatively trouble-free.

Electric Clocks

Did you ever notice how accurately an electric clock keeps time? Or have you ever wondered about the great confidence of electric clock-makers, in that they don't even provide "slow-fast" adjustment levers on their products?

There is a good reason for an electric clock's accuracy. Its synchronous motor runs at a dependably constant rate of speed. This is because a synchronous motor utilizes the alternations of house current to activate exactly 120 times each second (twice for each of the 60 cycles).

All it takes, then, is a proper gear system to move the hands of the clock exactly 1/60th of a turn each minute, or 1/12th of a turn each hour. The tiny motor, using just a fraction of a watt, turns a pinion gear. The pinion drives one gear, which drives another, and so on, in a complicated stepping-down train that finally reaches the clock hands.

Ordinarily, you need expect no problem with the gears. There is seldom a problem with the rest of the unit, either, except that a sudden power surge may sometimes burn out the fine wire of the motor coil. Many manufacturers have the coil so situated that it is simple to replace. Cheaper models may need replacement of the entire motor.

The workings of most clocks are easy to get at. A few screws or simply a snap-out "glass" usually bare the entire assembly. But you may have a problem getting at the gears, since many are sealed in. If so, chances are your problem is not within the gear system, anyway.

Clocks rarely need oiling. Sometimes, though, a noisy or buzzing movement may indicate worn motor-shaft bearings. A few drops of very light oil should stop this. Never use heavy oil anywhere in a clock. If there is dust or dirt in the movement, take off the case and dunk the innards into a container of kerosene. This will serve to both clean and lubricate the clock.

Just remember, though, that an electric clock's components, especially the hands and small gears, are delicate. Always be sure to take the greatest care when doing any work on this appliance.

Hair Dryers

The hair dryer can also be used to dry stockings, lingerie, even your face. And it is a dandy tool for defrosting a refrigerator or frozen foods. Some models have fingernail-drying and buffing devices as well.

The hair dryer has two basic parts: a motor-driven fan and a heating element. The fan blows air over the element and out through a nozzle. Up to this point, all dryers are pretty much the same. From the nozzle onward they divide into two types: hand-held and hood. Although these look radically different, their operation is almost identical. The difference is that, on the hood models, a length of plastic hose is attached to the nozzle, along with a hood that has a number of small openings to diffuse the warm air.

The unit has a dual switch that makes it possible to turn the motor on without the heater, but not vice versa. There is another switch, too — the thermostat. This shuts off the heating element when it gets too hot and turns it back on when it cools sufficiently.

Using your tester is the sure way to check for the cause of failure. If neither motor nor heater is working, it is a good bet that the trouble is in the cord assembly — at its common connection to both motor and heating element. If either of the two works and the other does not, start investigating the no-work area and let the working part alone. When you have located the source of trouble, replacement of the malfunctioning part is usually indicated unless the fault is simply a loose connection.

On hood types, you occasionally find a leaky hose. Plastic tape makes an effective patch.

Vaporizers

A vaporizer can be a life-saver when there's a sick person in the house, particularly a child with lung congestion. The typical unit uses two electrodes in water as a heating element. Electric current flows from one electrode to the other, causing the water to produce steam. A cup for medications is provided below the steam outlet. When the water level drops below the electrode ends, the unit shuts off.

This appliance normally gets infrequent use. Even so, it tends to build up heavy lime deposits. If these are not cleaned off periodically, you may find the vaporizer will not work just when you need it most. All visible deposits should be cleaned off immediately. Holes in the top and bottom of the unit should be cleaned out every time the vaporizer is used. After extended use, the housing should be removed and the electrodes cleaned. Do not lose the insulating spacer and be sure the electrodes are not touching.

Shopping for Replacements

If your appliance should develop malfunction, check the manufacturer's warranty before taking it to a dealer for repair. If the appliance is still covered by the warranty, manufacturer's repairs should cost you nothing.

There are many sources for replacement parts, yet frequently it can be difficult to find the exact part you need. The first place to check is the local dealer for the brand of appliance you have. If you don't know who he is, look him up in the Yellow Pages of your phone book under "Electric Appliances — Small — Retail" or "Electric Appliances — Small — Repairs and Parts." You might also find what you're looking for under "Electrical Repair Service," but this is mostly concerned with larger appliances. When you visit the dealer, take along the defective part.

If you can't find a part locally you may have to order what you need directly from the manufacturer. Make sure that you include the model number, year of purchase, and part number (if you know it) in your letter. Or send along the defective part and request that they mail a replacement to you C.O.D. Large retail and mail-order houses have Customer Service Departments that may also be of help. Furthermore, such major dealers often include a numbered parts list in the literature accompanying the appliance when you buy it. Never throw away such a list.

Modern built-in unit has electronic time-delay exit/entry feature —allowing family members to leave and reenter without setting off the alarm. Includes a fire and smoke detector.

8
Home Security Guide

Somewhere in the United States, a locked home is broken into every twenty seconds. City dwellings are the most common targets. Crime rates for cities are twice the national average, and the rate for burglary is six times higher in cities than rural areas. But suburban sections — the "outer cities" — fare little better than the inner core. Everywhere people are making sure they lock their doors. Still, ordinary locks are not enough. Most of them can be conquered in a matter of seconds.

Not surprisingly, the market for security devices is booming. Special locks, barriers, and alarms of one sort or another become instant best-sellers as fast as their inventors can patent them. "Security" today takes an endless variety of forms — including some that depend on electricity. These are of widespread interest among apartment dwellers and householders across the country. Hence, it is timely and helpful to include in this book a review of such protective electrical devices — though the emphasis here is on selection and installation rather than repair.

Of course, the decision to install any electric or electronic system depends on a multitude of factors.

What Are You Protecting?

This most obvious question would seem to have an obvious answer. Yet, ask it of anyone seeking protection and the reply may not be so clear-cut. The average homeowner may state that he wants an electronic security system in order to protect his family. Still, although in his mind the family is the ultimate "protectee," what he may really want is to protect his valuable possessions. A professional burglar, after all, rarely wants to bother the family. As a matter of fact, that's the last thing he wants to do.

Unfortunately, though, crimes of violence have increased sharply in recent years, even in what were formerly thought of as "quiet" or "nice" residential neighborhoods. All too frequently, horrifying headlines report the more grisly of these transgressions — break-in rapes, beatings, stabbings, etc. So security is no longer simply a matter of thwarting your classic burglar, who probably does not even carry a weapon. Real security requires protection against the

Early warning system for specific area. Pressure-sensitive doormat sets off a howling alarm when stepped on.

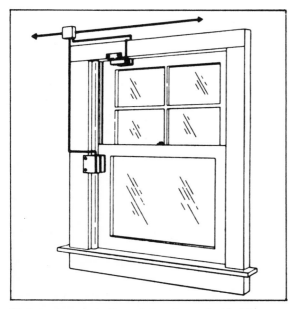

It's best to bug the bad guys before they get inside. This type of contact device will ring alarm bell as soon as double-hung window is disturbed.

psychopath bent on destruction and violence. It calls for sophisticated measures.

So when planning a working security system, it is necessary to keep in mind exactly *what* you want to protect. Only then can you decide *how*. For example, you could install a perimeter system all around your ¼-acre suburban estate to detect intruders as soon as they set foot on your lot. But unless you are trying to protect your prize roses or some priceless yew trees, this would be an expensive bit of "overkill." It would probably be better to install a system protecting against entry of the house itself, and simply build a high fence if all you want to do is keep out the neighbor's dog.

On the other hand, alarms around the doors and windows on the home's first floor will do nothing to deter a "second-story" man — and some foolish homeowners are so obliging that they leave ladders in the open or in unlocked garages for the convenience of such burglars. It should be remembered that an electronic system is to be used in conjunction with other forms of security — including common sense.

UL Approval

It is important that the electric security equipment you buy be UL-approved (see Chapter 3). When it was discovered that electric technology could provide a means of detecting and frightening away burglars, a tremendous variety of burglar alarms appeared on the market. Literally hundreds of manufacturers plunged into the field; some of them were legitimate and competent but many were just fly-by-nighters.

At first people bought these devices eagerly. Purchase was encouraged by insurance companies, who saw the alarms as a panacea for losses, which were mounting each year. Most theft contracts gave substantial discounts to homeowners whose premises were equipped with such burglary protection.

It soon became evident that the alarms were not perfect. Though some of the installations were effective, they tended to instill a false sense of confidence in many householders and their families. Furthermore, much of the equipment was plainly inadequate. Many of the heaviest losses were incurred at locations protected by "systems" that were in reality only simple open circuits like those used in doorbell wiring. As the true situation surfaced, the insurance companies abruptly reversed policy and eliminated automatic discounts for homes equipped with security devices.

A joint committee was then formed by insurance organizations, legitimate alarm companies, and UL. The committee set up minimum standards to apply to security alarm industry.

Only products meeting or exceeding these standards were to receive the UL-approved seal. It should be stressed that the standards dictated by UL represent the *minimum* acceptable requirements for properly designed alarm equipment. Product engineering and performance may often exceed these standards, and one system may be far superior to another in actual performance. Nevertheless, if systems do meet minimum standards they are designated UL-approved with no distinction as to which, if any, is a "better mousetrap."

Ways and Means

Electric security can be viewed from three standpoints: (1) type of system, (2) type of equipment, and (3) type of protection.

Take the familiar metallic foil and wiring setup seen on the windows and doors of many commercial buildings. As far as *type of system* is concerned, the installation is very likely a central station alarm system. This is connected by direct line to a local center — usually a private security company — which dispatches investigators as soon as an alarm rings. Usually, the police are notified as well. With this type of protection, all the potential entry points of a building are rigged so that an intruder will set off the alarm when he enters by window, door, or skylight.

Concerning the *type of equipment*, this installation utilizes electromechanical gadgetry. The mechanical act of breaking the tape or wiring causes a break in the electronic circuitry, sounding the alarm. But not all electromechanical alarms are perimeter protection; nor is all protection electromechanical. Various types of

systems provide the desired *type of protection*, using a variety of types of equipment.

Local Alarm System: In a local alarm system, the protective circuits are directly connected to a sound-producing apparatus (such as a bell or siren). The major disadvantage is that there are times and places when only the burglar will hear the bell. In an area where people are sure to respond, it is generally quite effective. The local alarm goes off as soon as the burglar enters the premises. It has the further advantage of scaring the burglar away before he can steal anything. If the alarm emits earsplitting wails, the psychological effect is enough to unnerve even the professional crook. The addition of blinding, flashing lights or other visual deterrents will also help the effectiveness of a local alarm system.

A local alarm system is well suited for most residential applications. The sounding device must be prominently located so that neighbors or police know where the noise is coming from. It must be on the exterior of the building, fully protected against weather and tampering, and it must be audible for a distance of at least 400 feet.

Central Station Alarm System: In a central station alarm system, the protected area has a direct connection to an alarm panel in a centrally located headquarters. The latter receives the alarm immediately via radio, leased telephone lines, or regular telephone lines. In the latter case, electronic devices usually have a programmed number or numbers automatically dialed when triggered by the protective device. The central office dispatches its investigator to the scene immediately, and the police are usually notified at the same time. Most systems of this type are dependent upon privately owned companies, and, therefore, are expensive. But there is no reason such a system cannot be operated by the community.

Proprietary Alarm: This type of installation is similar to a central station alarm, except that the alarm headquarters is on the premises. Obviously not for the average homeowner, the system is used most effectively on a large campus or industrial plant sites where there is an internal security force. A control panel is located in a constantly manned guard room and is connected to the security devices in the various buildings located on the premises.

Police Connection: Here a monitor is installed at the local police station and connected to the alarm system in one of the ways described above. There may also be backup links to local or proprietary systems.

The chief defect is that police may be swamped by such connections, most of which are subject to frequent false alarms. Like the boy who cried "wolf!" too often, families supposedly protected can count on a slow response, if any. In some localities, these systems have actually been outlawed because the many false alarms tie up too much police time.

Types of Protection

The sooner you catch a thief — or better, deter him — the safer you are. These goals are generally best achieved by installing detectors on exterior doors, windows, and sometimes walls and ceilings.

In addition, however, you may want to have a backup system in certain specific areas. Perhaps you don't have an original Dali or a priceless stamp collection, but you may have a safe or strongbox somewhere in the house containing money or valuables. A specific-area alarm device is vital in the event that a clever burglar is able to neutralize your overall alarm system.

This protection may take many forms. One of the simplest is a taut wire stretched around the protected area. So small it is almost invisible, the wire is connected to a snap-action switch that is activated whenever the tension is either too tight or too loose. The switch sounds an alarm when an intruder either brushes against the wire or cuts it. The wire used for such installations is specially treated to compensate for expansion and contraction due to weather changes.

Another type of specific-area protection is a pressure-sensitive doormat placed in the appropriate location. This will break a circuit if stepped on, setting off an alarm. A burglar may have skillfully cut a window around the metallic tape and crossed wires to nullify the system, but by stepping on the doormat, he will set off the alarm anyway.

Photoelectric, ultrasonic, audio, an radar installations are widely used in commercial and industrial establishments and are being adapted for residential use as well. Plug-in ultrasonic alarms — they look like small table model radios or large books — conceal systems that send out waves over large areas of a room. An intruder who steps into the area covered by the silent, harmless waves automatically activates an alarm. Photoelectric units can also be set up at doors, windows, or around the perimeters of a room. When the beam is broken, an alarm is tripped.

Automatic Telephone Dialers

This is a popular device used by homeowners, businessmen, and others who feel they cannot afford the more expensive central station and proprietary alarm systems. The dialer is connected to some type of intrusion alarm equipment and will automatically call a certain number, such as the nearest police station or a businessman's home, if the alarm is tripped. This does away with the need for constant surveillance by a central alarm company, the police, or others. When the number is called, a tape-recorded message will say something like: "A burglary is in progress at 210 Main Street. Please respond immediately."

This is an easy and relatively inexpensive way to alert the proper people to a break-in. But one of the big problems is the high nuisance or false-alarm rate. Police in many communities have been instructed not to respond to such calls. Some police departments have slapped a moratorium on further installations. There are, of course, answering services, neighbors, or a personal phone (if the alarm is at a place of business) that could be dialed in case of trouble. Some central station security services will allow the use of their phone numbers and will send their own men to respond to the call. As can be expected, this adds to the cost.

Telephone dialers are easy to install, since they use existing telephone lines. Some of the

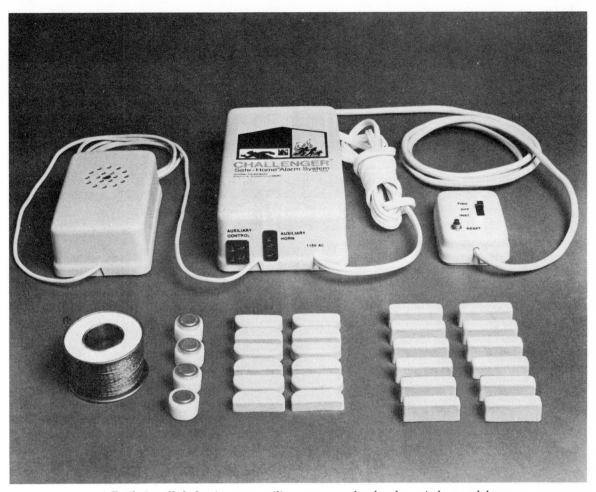

Easily installed plug-in system utilizes contacts to be placed on windows and doors.

more complex units have multiple channels that may be used to monitor for fires or watch for other problems, such as refrigerator failure, in addition to break-in and burglary. Each kind of occurrence activates its own individual channel and sends its recorded message to its own predetermined telephone number. The fire channel, for example, would be activated by a thermostat and would then alert the fire department.

But it is best not to depend entirely upon the taped message. A smart burglar might call the number and tie up the line, or he can call the answering number and tie that up while he loots your place (and the automatic dialer keeps getting that aggravating busy signal).

A main disadvantage of these devices concerns the lack of line supervision. There is always a chance that the call will not get through (even human dialers often face this problem). And a burglar may cut the phone wires as a precaution whether he suspects a dialer or not. Such difficulties can be avoided, but again the price rises, in some cases to prohibitive levels.

Unskilled installers and poor maintenance of electrical systems may cause such a rash of nuisance alarms to local police that the dialer calls are ignored or even outlawed — as mentioned above — by local police.

Finally, the dialer method is not the fastest way to catch a thief. Consider the steps that the dialer goes through before help is forthcoming:

1. After the intrusion alarm is tripped, the dialer is alerted and must connect itself to the telephone line.
2. The dial tone has to start. At times, this delay can be substantial.
3. After the dial tone sounds, the number must be dialed. It may not seem to take long to dial a phone number, but try it sometime when you are in a hurry! Each digit actually interrupts a circuit a certain number of times — "1" makes one interruption, "9" interrupts the circuit nine times. Touch-tone dialing, if you have it in your area, helps somewhat. Tone phones operate by utilizing different sounds once each instead of numbers.
4. There is another delay while the phone is answered at the other end of the line, or when a busy signal is heard.
5. The message must be relayed to someone at the other end.
6. The person at the other end has to take some action, either by coming to the scene or by notifying the authorities, causing the loss of more precious minutes.

The disadvantages of a piece of equipment do not mean that it is of no value. But they should be recognized and steps should be taken to counteract them. In most instances, the remedy is to utilize more than one form of protection.

Lock-Alarms

Especially popular with apartment dwellers (but equally effective for homes, offices, or small businesses) is the combination lock-alarm. A typical unit consists of a dead-bolt lock and chain coupled to a highly sensitive, battery-powered, solid state electric alarm. Any attempt to pick the lock cylinder or to force the lock, dead-bolt, chain, or door itself sets off a shrill, pulsating audio signal to provide a psychological deterrent to breaking and entering.

In operation, the dead-bolt performs in a conventional manner in that it can be engaged or disengaged by turning a knob from the inside or by using a key from the outside. With

the chain alone in place, an occupant may pre-screen visitors by opening the door slightly. Any attempt by a person on the outside to force the door with only the chain engaged also sets off the alarm.

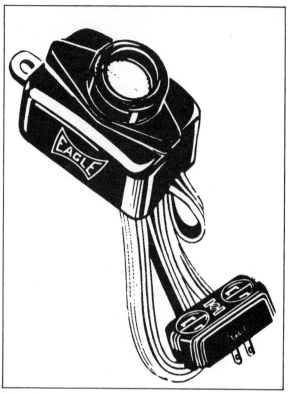

Photoelectric cell automatically turns off house lights at daybreak and turns them on at nightfall. Scares off burglars while you are away on vacation or at the movies.

Protection for Your Home

There are enough electric gadgets around to protect even the most vulnerable home. Unfortunately, the cheaper equipment is not wholly foolproof, and total protection is out of the reach of most middle-income families — who bear the brunt of burglaries.

The proliferation of police-connected devices

RIGHT: Complete AM-FM radio/intercom/intruder-detection system does everything but wash the dishes. It also has jacks to accommodate a record player and a fire-warning system. If the music doesn't scare away crooks, a shrieking alarm activated by sensors on doors and windows will.

is sure, ultimately, to turn the police against such equipment entirely. Local alarms are unreliable, since they depend on neighbors who care.

What the future probably will see is the development of small central alarm systems, perhaps run by the community or by neighborhood cooperatives. There are certain to be more innovations in the field of home security, because the demand is great.

In the meantime, you can definitely achieve a reasonable level of protection for your home.

First, carefully evaluate your needs. Then install the most appropriate electronic alarm or lock system, or both. Coupled with personal vigilance and common sense on the part of your family and yourself, these systems should keep your home safe.

Some systems can be installed by a handy do-it-yourselfer. Others require professional expertise. A reputable dealer in alarm devices can steer you right — let him advise you as to whether or not you should tackle the installation yourself.

Electrical Terms at a Glance

AC: Alternating current, the kind serving most homes. Usually the alternations occur at a rate of 60 cycles per second but 25- and 50-cycle systems are used in a few localities.

Ampere: Unit for measuring the quantity of electricity passing through a circuit at any given instant. Often called "amp."

Ballast: Coil that controls voltage flow in fluorescent lights.

Box: Standardized boxlike housing that protects wire connections from mechanical damage.

Box clamp: A fitting that attaches cable to boxes.

Branch circuit: All devices and wiring controlled by a single fuse or circuit breaker in the service entrance.

Bushing: An insulator that fits around ends of BX cable to prevent cut metal from fraying the wires within.

BX: Common trade term for flexible armored cable.

Cable: Lengths of two or more heavily insulated wires enveloped in a tough sheath.

Canopy: Portion of a light fixture that contains its connections to a box.

Circuit: The wires and devices through which electric current travels from a point of supply and back to it.

Circuit breaker: A switch that performs the same function as a fuse by opening or "breaking" a defective circuit.

Code: Term for the National Electric Code de- vised by the National Board of Fire Underwriters. The code sets standards for materials and methods used in electrical systems. Term also applies to particular regulations in force in any locality.

Conductors: Trade jargon for "electric wires"; more generally, any substance or body capable of transmitting electricity, including yourself.

Conduit: Metal tubing through which wires are run for protection.

Connector: A fitting to secure cable to a box; substitutes for a box clamp.

Cord: Flexible length of double wire for connecting a lamp or appliance to an outlet.

DC: Direct current, still being used in a few areas. As it does not alternate in cycles, it will not operate some kinds of appliances.

Fish wire: Narrow, springy metal tape bent into a hook at one or both ends. Used to pull wire through walls, floors, and ceilings in existing homes.

Fuse: A safety device that breaks the flow of current when a circuit is shorted or overloaded.

Fuse box: A type of service entrance that contains fuses.

Greenfield: A flexible type of conduit.

Grounding: The connection of an electrical system to the earth. This precaution minimizes damage from lightning — and prevents electric shocks.

Ground wire: See *Neutral wire* below.

Heater cord: Cord insulated against heat transmission to the user or surrounding surfaces. Used for appliances operating at high temperatures, such as electric irons and portable space heaters.

Hickey: Device used to mount a hanging light fixture to a ceiling.

Hot wire: The wire, usually colored black or red, that carries current from the point of origin to point of use.

HP: Abbreviation for horsepower, the unit for measuring work rate. One HP equals 745.7 watts or, in mechanical terms, 33,000 foot-pounds of work per minute.

Insulation: Sheathing over wires to prevent the escape of electricity. Also, any nonconductive material; that is, material resistant to the passage of electricity or heat.

Jumper: A length of wire used to extend another wire during testing; for example, to extend a tester probe to a terminal otherwise out of reach. Term also is applied to any short-length wire installed to bypass a circuit or used temporarily to close a circuit.

Junction box: Type of protective box in which wires of different circuits are connected.

Knockout: Portion of box easily removable to provide an entrance for cable.

Light box: Type of protective box in which wires of light fixtures are connected to circuit wires.

Line: Cable comprising or supplying a circuit.

Line cord: Same as *cord*.

Live wire: Same as *hot wire*.

Mercury switch: Switch that operates noiselessly because an internal flow of mercury makes and breaks the contacts.

Neutral wire: The wire, usually colored white or pale gray, that carries current from point of use to point of origin. Also called *ground wire*, because the neutral wires of an alternating current system must be connected to ground.

Ohm: Unit for measuring resistance to current flow.

Outlet: A device for tapping the electric current in a branch circuit. The term is sometimes used synonymously for *receptacle* or *light box*.

Outlet box: Type of protective box in which wires of an outlet are connected to circuit wires. Term is sometimes used synonymously for *light box*.

Overload: More current in a circuit than it is designed to carry. Results are overheated wiring and blown fuses.

Parallel cord: Type of light-duty lamp cord not covered by a sheath. The two wires of the cord have insulation that holds them parallel, but they are more easily separated than the wires of sheathed cord.

Quiet switch: A switch much less noisy than customary household switches but not as completely noiseless as a mercury switch.

Receptacle: Female portion of an outlet or female plug of a heavy-duty cord. Also used synonymously for *outlet*.

Resistance: The opposition of any material to current flow within it. Resistance converts electrical energy into heat energy.

Romex: Trade name for nonmetallic sheathed cable.

Running board: Any board to which cable is strapped or stapled for support.

Service entrance: The protective box in which current supplied by a utility company is distributed among the branch circuits of a home. The term is frequently extended to include the meter, external connections, and other utility-supplied portions of the electrical system.

Service panel: Same as service entrance. Also, the panel within the service entrance that carries wiring, switches, fuses, or circuit breakers.

Short circuit: Accidental or inadvertent electrical connection between a hot wire and a neutral or ground wire. The route of the connection may be through adjacent bare metal.

Solderless connector: A device that splices and insulates wires to make a safe electrical connection.

Starter: An automatic switch initiating and maintaining current flow in a fluorescent light.

Switch: A device for breaking current flow and restoring it.

Switch box: Type of protective box in which switch terminals are connected to the wires of lights or other devices.

Switch loop: The hot wire delivering current from a device to a switch plus a second hot wire returning the current to the device. Switch loops should never be connected to neutral or ground wires. If the white wire in a cable is used to complete a switch loop, that wire should be painted black or red to indicate that it is hot.

Takeoff: The wiring, connections, box, and other portions of a permanently installed tap.

Tap: Any wiring installed to draw current from a circuit.

Three-way switch: A kind of switch used in pairs to control a device from two different points.

Underwriters' Laboratories: A nationally recognized organization that tests wiring materials and electrical devices to make certain they meet minimum standards for safety. If approved, an item carries a UL seal or tag.

Volt: Unit for measuring the force or "pressure" with which current flows through a circuit; something like pounds of pressure per square inch in a water system.

Voltage drop: Term describing loss of voltage that occurs when current passes through wire. The drop is due to the internal resistance of the wire. The longer a wire, and the smaller its cross-section, the greater voltage drop.

Watt: Unit for measuring electrical power being delivered by a circuit. One watt is equivalent to the work rate of one ampere at a pressure of one volt (a volt-ampere).

Watt-hour: One watt used for one hour equals 1 watt-hour; 1000 watt-hours equals 1 kilowatt-hour (kwh), which is the unit by which electricity is metered and billed you by the utility company.

Index